Potato Diseases

Potato Diseases

AVERY E. RICH

Department of Botany and Plant Pathology
University of New Hampshire
Durham, New Hampshire

1983

ACADEMIC PRESS

A Subsidiary of Harcourt Brace Jovanovich, Publishers

New York   London
Paris   San Diego   San Francisco   São Paulo   Sydney   Tokyo   Toronto

ACADEMIC PRESS, INC.
111 Fifth Avenue, New York, New York 10003

United Kingdom Edition published by
ACADEMIC PRESS, INC. (LONDON) LTD.
24/28 Oval Road, London NW1 7DX

Library of Congress Cataloging in Publication Data

Rich, Avery E.
    Potato diseases.

    Includes bibliographical references.
    1. Potatoes--Diseases and pests.   I. Title.
SB608.P8R43 1983   635'.219   82-24290
ISBN 0-12-587420-0

PRINTED IN THE UNITED STATES OF AMERICA

83 84 85 86   9 8 7 6 5 4 3 2 1

To the memory of my father, the late Nathan Harold Rich (1881–1960), who was a pioneer in the Maine Certified seed potato industry. He taught me a great deal about potato production and potato diseases from a practical standpoint and undoubtedly influenced me to become a plant pathologist.

Contents

4 Diseases Caused by Viruses, Viroids, and Mycoplasmas

5 Diseases Caused by Nematodes and Insects

6 Noninfectious Diseases

7 Seed Potato Certification

Glossary

Bibliography

Preface

This book should serve as a reference source and textbook for researchers, extension workers, teachers, students, and growers who are interested in the general subject of potato diseases. It is a culmination of a lifetime of practical and scientific experience in seed potato production and the numerous diseases encountered.

It includes chapters on bacterial and fungal diseases, diseases caused by viruses, viroids, and mycoplasmas and by nematodes and insects, noninfectious diseases, and seed potato certification. For each of the major diseases, their names, importance, causal agent, etiology, and control are discussed.

It is hoped that this book will serve a useful purpose for both the scientific community and the layman.

Avery E. Rich

Acknowledgments

I wish to express my deep appreciation to all those generous people who supplied me with photographs illustrating some of the potato diseases. Special thanks go to Dr. Frank Manzer and Dr. Richard McCrum of the Maine Life Sciences and Agriculture Experiment Station, to Dr. William Mai of Cornell University, and to Dr. James Bowman of the University of New Hampshire. I wish to acknowledge the willing and efficient assistance of Ms. Suzanne Lefevre in typing the manuscript. Finally, I want to thank my wife, Erma, for her cooperation, patience, and understanding during the preparation of this book.

Disclaimer

Due to the fact that the use of most fungicides, insecticides, nematicides, and herbicides is regulated by state, provincial, or federal regulations, and the fact that these regulations are undergoing constant change, the author and the publisher assume no responsibility for the improper or illegal use of any chemicals on potatoes. Growers should check with their Environmental Protection Agency and their county, state, or federal Extension Service for the latest recommendations and label approval of chemicals for control of potato pests.

1

Introduction

The Irish potato (*Solanum tuberosum* L.) is an important world food crop. Apparently it originated in the Andes Mountains of South America. According to Scott (1976), the Inca Indians first cultivated the potato in 200 B.C. The Spanish conquistadors found the Incas growing the potato, which they called *papa,* in 1537. They introduced it into Spain from where it spread into Italy and then into central Europe. Gradually it became a staple food crop in Europe, especially in Germany, Russia, and Ireland. The potato was introduced from Ireland into the United States in 1719 where it was first grown in Londonderry, New Hampshire (Smith, 1968). Hawkes (1978) has written an excellent history of the potato.

The geographic range of the potato is almost worldwide. It is grown as a major source of food in most countries with a temperate climate.

The potato is susceptible to a great many diseases, some of which are widespread and others are localized. The causal agents of these diseases include bacteria, fungi, viruses, mycoplasmas, viroids, and nematodes. Another group of disorders, called noninfectious diseases, include those due to unfavorable environment, faulty nutrition, or other abiotic factors (Dykstra, 1948; Dykstra and Reid, 1956; Houghland *et al.,* 1957; Kehr *et al.,* 1964; Whitehead *et al.* 1953).

The "Index of Plant Diseases in the United States" (Anonymous, 1960) lists approximately 160 diseases and disorders of *Solanum tuberosum.* About 50 are caused by fungi, 30 by viruses, 10 by bacteria, and another 50 or so are either nonparasitic or due to unknown causes. Several others are due to nematodes or insects, and one is caused by dodder. Conners

(1967) lists 85 potato diseases in Canada. He includes both the English and French name for each disease. Blodgett and Rich (1950) described and illustrated 68 diseases and defects of potato tubers in the Pacific Northwest, and Folsom *et al.* (1955) recorded about 60 diseases of potato in Maine. O'Brien and Rich (1976) recently described 61 diseases of potato plants and/or tubers in the United States. Lapwood and Hide (1971) discussed the important nonvirus diseases of potatoes in the British Isles, and Harrison (1971) discussed potatoes viruses in Britain.

Wellman (1972) states that diseases are more numerous in the tropics than in the temperate zone. He reported that *Solanum andigenum - tuberosum* has 175 diseases in the tropics compared to 91 in the temperate zone, but he did not identify them. Smith (1972) mentions at least 17 viruses and two mycoplasmas which affect potatoes. A rather complete compendium of potato diseases was published in 1981 (Hooker, 1981).

The Irish Famine, due to late blight caused by *Phytophthora infestans,* was one of the most dramatic episodes caused by a plant disease (Large, 1940). The Irish depended almost entirely on potatoes as their staple food. When the blight practically wiped out their potato crops during the middle of the nineteenth century, one million people died in Ireland and another million and a half emigrated to the United States and other countries (Western, 1971).

Late blight is an example of a disease which can ruin the crop. Other diseases, such as common scab, never cause crop failures but may severely affect the marketability or keeping quality of the crop. Great strides have been made in the control of potato diseases through the use of effective fungicides and insecticides, improved cultural practices, and the development of resistant cultivars. However, considerable sums of money are spent each year to effect these controls. In addition, most resistant cultivars are resistant to only one or two diseases and are susceptible to most of the other diseases. Therefore, the battle of disease control is far from over.

Potatoes are usually clonally propagated by planting potato seed pieces or sets, or whole tubers. This method of propagation lends itself inevitably to the introduction of potato pathogens from one area to another and to the overwintering of pathogens in the stored tubers which may be used for propagative purposes. Systems of certification or registration of seed potatoes have been developed in an attempt to minimize these dangers, but they have been only partially effective. In some cases, quarantines have been used to control the spread of dangerous pathogens.

An attempt will be made to group the potato diseases into chapters based on their causal agents. In most cases, each disease will be discussed from the standpoint of names of the disease, importance, causal agent, symptomatology (symptoms and signs), epidemiology, and control.

2

Bacterial Diseases

I. INTRODUCTION

The bacteria which attack potatoes are a comparatively small but important group of pathogens. Bacteria are single-celled, microscopic organisms which reproduce by fission. They belong to the class Schizomycetes, sometimes referred to as the fission fungi. All are non-spore-forming short rods in the order Eubacteriales with the exception of *Streptomyces scabies,* the causal agent of common scab (Dowson, 1957; Elliott, 1951). All are gram-negative except *Corynebacterium sepedonicum,* the causal agent of ring rot.

Streptomyces is often classified (Anonymous, 1960; Drechsler, 1919) with the fungi rather than with the bacteria. It produces rudimentary hyphae which produce spores by the formation of septations. It is also (Breed *et al.,* 1957; Rich, 1968; Walker, 1969) classified with the bacteria. It belongs to the order Actinomycetales and family Streptomycetaceae.

Bacteria are primarily wound parasites, but they frequently enter uninjured plants through natural openings such as stomata and lenticels. Insects are sometimes involved in their dissemination and may make wounds in healthy plant tissues through which the bacteria gain entrance.

Bacterial growth is favored by warm, moist conditions. Bacteria thrive on media containing starches and sugars. Thus potato tubers are a good medium for their growth.

Some bacteria attack only tubers, while others attack both tubers and plants. The following diseases will be discussed here: bacterial soft rot, black

3

leg, brown rot, common scab, pink eye, ring rot, and miscellaneous bacterial diseases.

II. BACTERIAL SOFT ROT

Bacterial soft rot is a common disease of many fruits and vegetables, including carrots, onions, and potatoes. Frequently it is preceded by some other disease or injury and takes over during storage or transit of the harvested crop.

The disease is called bacterial soft rot or simply soft rot in English. The French name for it is *pourriture molle bactériene;* the Germans call it *Knollennässfaule* (Miller and Pollard, 1976).

A. Importance

Walker (1969) considers bacterial soft rot one of the most destructive diseases of vegetables in storage and transit. It is worldwide in its distribution (Chupp and Sherf, 1960). Cromarty and Easton (1973) reported that 7.9% of rail cars of Washington State potatoes are rejected due to defects and that 63% (5% of cars) was due to tuber decay. Washing potatoes in deep vats or dirty water also increases soft rot (Dewey and Barger, 1948), as does exhaustion of oxygen and accumulation of carbon dioxide in the air surrounding potato tubers (Nielsen, 1968). Exposure to solar radiation for 60 min, which raises the tuber temperature to 45°C, also increases susceptibility to soft rot (Nielsen, 1954).

B. Causal Agents

The organism primarily responsible for bacterial soft rot is usually classified as *Erwinia carotovora* (L. R. Jones) Holland. Jones originally named it *Bacillus carotovorus* in 1901. *Erwinia aroideae* (Towns.) Holland is also credited with causing soft rot of several vegetables and iris corms. There appears to be considerable disagreement concerning *E. atroseptica* (van Hall) Jennison, the role it plays in the soft rot and blackleg syndromes, and whether or not it is a distinct species or only a subspecies of *E. carotovora* (Burkholder and Smith, 1949; Graham, 1964; Smith and Ramsey, 1947; Walker, 1969). (See Blackleg, Section III, this chapter.)

Erwinia carotovora is a non-spore-forming, gram-negative, short rod, 1.5–3.0 μm long by 0.6–0.9 μm wide, with rounded ends, occurring singly or in chains. It is motile with two to six peritrichous flagella. In culture the minimum, optimum, and maximum temperatures for growth are 2°, 25°, and 37°C, respectively. Disease development occurs between 5° and 37°C, with an optimum of 22°C. A temperature of about 50°C kills the bacteria (Agrios, 1969). *Erwinia aroideae* differs from *E. carotovora* and *E. atroseptica* by its lack of gas formation on certain carbohydrate media (Walker, 1969).

C. Symptomatology

Symptoms on potato are usually, but not always, confined to tubers, and usually develop during storage or during transit. Infected tubers typically develop a watery soft rot accompanied by an offensive odor. Small areas around wounds or the entire tuber may be affected (Fig. 2.1). If the bacteria invade the lenticels, numerous small sunken areas, 0.3 to 0.6 cm in diameter, develop more or less over the entire surface of the tuber. Affected tuber tissue is typically white to cream colored. It is soft and watery under moist conditions but may be chalky-white if the humidity of the air sur-

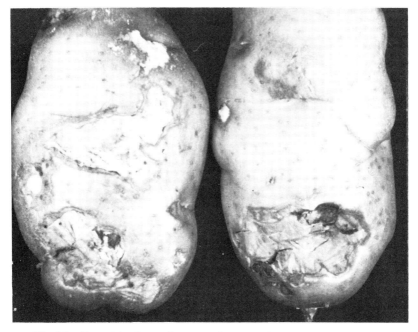

Fig. 2.1. Bacterial soft rot of potato tubers caused by *Erwinia carotovora.*

rounding the tubers is considerably reduced. A clear, amber liquid may exude from decayed areas, or in severe cases the tubers may collapse (O'Brien and Rich, 1976; Ruehle, 1940). Enzymes produced by the bacteria dissolve the middle lamella of the host tissue, thus loosening the cells from one another and causing tissue breakdown.

Davidson (1948) described a soft rot of potato tuber initials (small newly formed tubers) in wet soil. He isolated *E. carotovora, E. aroideae,* and two *Bacillus* species from infected lenticels. Erinle (1975b) recovered *E. carotovora* from potato stems which did not turn black but which developed a fast soft rotting or hollowing out of the stems. Presence of the bacteria is indicative of the disease.

D. Epidemiology

Bacterial soft rot is favored by warm temperatures and high moisture levels in the soil or during storage (Bennett, 1946; Cromarty and Easton, 1973; Davidson, 1948; Erinle, 1975b). Immature tubers are more subject to rot than are mature ones (Bennett, 1946; Smith and Ramsey, 1947). Wounds also favor the development of soft rot (Walker, 1969). Insects, especially seed corn maggots *(Hylemyia* spp.), make wounds and transmit the bacteria from one tuber or plant to another (Agrios, 1969).

Other predisposing factors are low temperatures, which freeze the tissue, bruises, and insect injuries (Walker, 1969). Washing potatoes in deep water vats for long periods of time also increases the chances of soft rot development (Dewey and Barger, 1948).

E. Control

Sanitation is imperative for control of bacterial soft rot. Warehouses should be cleaned and disinfested with a good disinfestant such as formaldehyde or copper sulfate.

Potatoes should be grown in well-drained soils. Heavy soils, wet spots, and excessive irrigation should be avoided. Only mature tubers should be harvested, if possible. Vine killing before harvest will hasten maturity. Harvesting in cool, dry weather will reduce sunscald, thereby decreasing the chance of soft rot development.

Careful handling to reduce bruising is essential. Potatoes should be stored or shipped under cool (15°–18°C), relatively dry conditions. Cooling boxcars with ice or cool air is beneficial. Air circulation will prevent exhaustion of oxygen and accumulation of harmful levels of carbon dioxide. Freezing in the field, storage, or during transit should be avoided.

Potatoes should not be washed before marketing unless necessary. If they are washed, they should be washed in clean water with spray jets and dried rapidly. Soaking in deep vats should be avoided.

Highly susceptible cultivars, such as Irish Cobbler, Cayuga, Pawnee, Teton, and White Rose should be avoided. Bliss Triumph, Essex, Katahdin, Kennebec, and Sebago are somewhat resistant to this disease (Nielsen, 1954; O'Brien and Rich, 1976).

Treating cut seed pieces with streptomycin reduced bacterial seedpiece decay but increased Fusarium dry rot (Bonde and Malcolmson, 1956). Rich *et al.* (1960) concluded that it did more harm than good.

III. BLACKLEG

Blackleg is a serious disease of potato plants, tubers, and seed pieces. It is called *jambe noire* in French, *pierna negra* or *pudrición suave del fruto* in Spanish, and *Schwarzbeinigkeit* in German. It occurs in Africa, Asia, Australia, Europe, North America, Central America, and South America (Miller and Pollard, 1976).

A. Importance

Blackleg is a serious problem during wet seasons and in irrigated fields. It is often the cause of a field being rejected for seed certification. In the United States, the author has occasionally observed fields with as much as 10% of the plants infected. The Sebago cultivar is highly susceptible. Burkholder and Smith (1949) reported that in New York it is transmitted to 6-7% of all tubers, thus requiring its inclusion in the seed potato certification program. Graham (1962) noted that blackleg is widespread in Scotland and may cause up to 30 or even 50% infection. It is transmitted in Scottish seed to warm countries such as South Africa, Southern Rhodesia, and Israel.

B. Causal Agents

There is almost total agreement that the primary causal agent is a bacterium in the genus *Erwinia* or *Pectobacterium*. However, there is considerable difference of opinion concerning the correct nomenclature for the species. Some authorities consider it to be a subspecies or variety of *Erwinia carotovora*, referring to the common soft rot organism as *Erwinia carotovora* var. *carotovora* and the blackleg organism as *Erwinia carotovora* var. *atroseptica* (Erinle, 1975a,b; Molina *et al.,* 1974; Tanii and Akai, 1975). Others consider it to be a species distinct from *E. carotovora* and refer to it as *Erwinia atroseptica* (van Hall) Jennison (Burkholder and Smith, 1949; Conners, 1967; Smith, 1949; Rich, 1968; Walker, 1969). The United States Department of Agriculture workers generally classify it as *Erwinia phytophthora* (Appel) Holland (Anonymous, 1960; Miller and Pollard, 1976; Malcolmson, 1959; O'Brien and Rich, 1976). Hellmers (1959) considers *Pectobacterium carotovorum* var. *atrosepticum* (van Hall) Dowson to be the correct name of the potato blackleg pathogen, as do Graham and Harper (1966).

The species of *Erwinia* are morphologically indistinguishable. However, several physiological and nutritional differences can be used to separate them. *Erwinia carotovora* flagella stain more readily with Casares-Gils' stain than do flagella of *E. atroseptica*. *Erwinia carotovora* utilizes ethyl alcohol, dulcitol, sodium hippurate, sodium malonate, and sodium urate, whereas *E. atroseptica* does not. *Erwinia atroseptica* has a slightly lower temperature range, producing more acid in maltose and less in glycerol than does *E. carotovora* (Burkholder and Smith, 1949; Smith, 1949; Walker, 1969). Differences in the species have also been suggested by serological tests (Malcolmson, 1959; Shenider and Murzakova, 1964).

Erwinia atroseptica incites blackleg, but *E. carotovora* usually does not (Burkholder and Smith, 1949; Smith, 1949). When separating the organisms on the basis of induction of blackleg, Erinle (1975a) found that inoculation of tubers was more reliable than inoculation of stems. Stapp (1947) recognized pathogenic strains of the blackleg pathogen, and Vruggink and Maas-Gusteranus (1975) can readily distinguish the two organisms using serological techniques.

Recently, Stanghellini and Meneley (1975) and Tanii and Akai (1975) reported the occurrence of blackleg caused by *E. carotovora* var. *carotovora* in Arizona and Japan, respectively. Thus the correct taxonomy and nomenclature of the causal organism or organisms are still unclear. However, most workers consider the causative agent of soft rot as *E. carotovora* var. *carotovora* and *E. carotovora* var. *atroseptica* as the causative agent of blackleg. They are gram-negative, motile short rods, 1.5–3.0 μm long by 0.6–0.9 μm wide.

C. Symptomatology

Infected plants develop characteristic symptoms. The leaves turn yellow and roll upward when plants are relatively small. Plants tend to stand upright. At first the underground portion of the stem turns black, but as the disease progresses, the inky black color advances up the stem for several inches. The stem may become slimy. Severely affected plants wilt and die. The disease also progresses downward through the stolons and into the tubers. A conical dark-colored rot, at the stem end of affected tubers is typical (Fig. 2.2) (O'Brien and Rich, 1976; Rich, 1968, 1977). The presence of *Erwinia atroseptica* is indicative of the disease.

D. Epidemiology

The pathogen overwinters in infested soil and infected tubers. It can be spread from diseased to healthy tubers by a contaminated cutting knife.

Insects, especially seed corn maggots [*Hylemya platura* (Meigen) and *H. florilega* (Zetterstedt)], spread the bacteria from diseased to healthy seed pieces. The bacteria are carried in the intestinal tract of the insects and are transmitted to healthy seed pieces when contaminated larvae, pupae, and adults feed on the cut surfaces of healthy seed pieces (O'Brien and Rich, 1976). The pathogen is also disseminated by fruit flies *(Drosophila melanogaster* Meig.) (Molina *et al.,* 1974).

Conditions which are unfavorable for rapid healing of freshly cut seed pieces favor disease development. Planting susceptible cultivars in wet soil predispose them to disease development. Certain cultivars, such as Sebago, Fundy, and Huron, are more susceptible to disease than others (Hodgson *et al.,* 1973).

Graham and Harrison (1975) demonstrated that the pathogen can be disseminated from plant to plant by simulated rain drops and concluded that rain and air currents are potential methods of spreading the pathogen.

E. Control

One of the most effective controls for blackleg is to plant only healthy, whole seed potatoes. If seed potatoes are cut, they should be suberized before planting. Seed potatoes should be warmed to about 12°–15°C (55°–60°F) before cutting. The cut seed pieces should then be stored for about 2 days at the above temperature and at relatively high humidity, with a free flow of air around and through them to prevent heating. Planting in cold wet soils which delay suberization and emergence should be avoided.

The maintenance of sanitary conditions is important to minimize inoculum formation and infestation by seed corn maggots and fruit flies. Crop rotations may be helpful in reducing soil inoculum.

Where feasible, resistant cultivars, such as Katahdin and Russet Burbank, should be planted. Cherokee, Hunter, Green Mountain, Irish Cobbler, Kennebec, and Pontiac are moderately resistant, whereas highly susceptible cultivars, such as Arran Consul, Fundy, Huron, Norgold, Sebago, and Russet Sebago (Hodgson *et al.,* 1973; Munro, 1975; O'Brien and Rich, 1976; Rich, 1968–1977), should be avoided.

Fig. 2.2. Potato plant exhibiting symptoms of black leg, a bacterial disease caused by *Erwinia atroseptica*. (Photo courtesy Maine Life Sciences and Agricultural Experiment Station.)

Stapp (1947) rated Carnea, Flava, Johanna, Priska, Robusta, Sickingen, and Stärkeragis as resistant. He placed Ackersegen, Alpha, Aquila, Centa, Depesche, Edelgard, Frühgold, Frühnudel, Krebsfeste, Kaiserkrone, Möwe, Parnassia, Pepo, Sandnudel, Speisegold, Tiger, Voran, Weisses Rössl, and Weltwunder in the moderately resistant category.

Seed treatment with antibiotics was suggested by Bonde and Malcolmson (1956), but further studies indicated that it may do more harm than good (Rich *et al.,* 1960). Graham and Harper (1966) found that increased applications of nitrogen or complete fertilizer reduced blackleg stem infection.

IV. BROWN ROT

Brown rot is also referred to as bacterial wilt, southern bacterial wilt, or slime disease. In New Zealand it is called mattery eye (Boesewinkel, 1976). French names for this disease include *bactériose vasculaire* and *pourriture brune.* It is referred to as *podredumbre anular, dormidera, marchitez bacterial,* and *enfermedad de limo* in Spanish. German names for this disease are *Braunfäule, bacterielle Wilke,* and *Schleimkrankheit* (Miller and Pollard, 1976). It occurs in Africa, Asia, Europe, North America, Central America, and South America, and was first reported in the United States by E. F. Smith in 1896.

A. Importance

Brown rot is an extremely destructive disease of potatoes in tropical and subtropical regions. It causes heavy losses in the South Atlantic and Gulf Coast states of the United States (Eddins, 1936; Kelman, 1953; O'Brien and Rich, 1976). It is not, however, a problem in the northern states and does not occur in Canada (Munro, 1975).

B. Causal Agent

The bacterium responsible for this disease is *Pseudomonas solanacearum* (E. F. Sm.) E. F. Sm. It is a gram-negative short rod, 1.5 × 0.5 μm, and is monotrichous (motile by a single polar flagellum). Colonies on agar are opalescent which become darker with age. They are small, irregular, smooth, wet, and shiny. The optimum temperature is 35°–37°C (Weber, 1973). There are at least three pathogenic races of *P. solanacearum:* race 1 attacks solanaceous crops such as eggplant, tobacco, tomato, and potato;

race 2 is parasitic on banana; and race 3 is highly pathogenic on potato but only weakly pathogenic on tobacco.

C. Symptomatology

The first symptom of the disease in the plant is a slight wilting and droping of foliage at the end of the branches during the heat of the day. Affected leaves become pale green and then yellow or bronze. Wilting becomes more severe each day until plants are permanently wilted, which can result in death. The vascular bundles in the stem turn brown and the stems become streaked.

The first symptom of the disease in tubers is a browning of the vascular ring near the stem end, which progresses until a slimy ooze develops at the eyes and stem end of severely affected tubers (Fig. 2.3). Soil adheres to the surface of the tubers where the slimy ooze has escaped. These tubers will continue to break down in the soil, in transit, and in storage (Weber, 1973). Secondary rot organisms may invade infected tubers, causing them to become a slimy mass with an offensive odor. Diseased plants may produce both healthy and infected tubers. Symptomless plants may also produce diseased tubers.

Gram-negative, short rod bacteria which plug the vessels of infected plants, thus causing the wilting, are signs of the disease. They are also present in large numbers in the tubers and are responsible for slimy ooze and breakdown (O'Brien and Rich, 1976).

D. Epidemiology

Pseudomonas solanacearum is a warm temperature organism with an optimum of 35°–37°C, which is also favored by a relatively high soil moisture level. It attacks other solanaceous plants besides potatoes, causing a serious disease of tobacco called Granville wilt. It thrives in slightly acid, neutral, and alkaline soils. It persists in most soil types including sandy, loam, clay, muck, and peat soils, but not in marl. Most potato cultivars are susceptible to infection by this organism. It overwinters in infected seed potatoes, infested soil, and solanaceous weed hosts. Invasion of the host plants occurs primarily through wounds. Nematodes and mechanical injury to roots by cultivation equipment aid root penetration. The disease may be introduced into new areas by infected seed tubers or infected tomato, pepper, or eggplant transplants.

Fig. 2.3. Potato plant exhibiting brown rot symptoms, a bacterial disease caused by *Pseudomonas solanacearum*.

E. Control

Where this disease is a problem, losses can be reduced by planting disease-free seed potatoes in noninfested soil. If the soil is infested, crop rotation and control of solanaceous weed suscepts must be practiced.

The addition of sulfur to sandy soils is helpful in controlling this disease. An application of 800 kg/ha (800 lbs/acre) followed by 3000 kg/ha (3000 lbs/acre) of dolomitic limestone in summer is suggested. Also useful is the addition of sufficient sulfur to sandy soils to lower the pH to 4.0, followed by the addition of an equivalent amount of lime (Eddins, 1936, 1939).

Since Chippewa, Green Mountain, Katahdin, Prisca, and Sebago cultivars are moderately resistant to brown rot, choosing one of these cultivars
for planting should help reduce losses from this disease (Nielsen and Haynes,
1960; O'Brien and Rich, 1976; Walker, 1953).

V. COMMON SCAB

This disease is frequently referred to only as scab. It is often called common scab to differentiate it from powdery scab, caused by the fungus
Spongospora subterranea (Walker) Lagerh. It is referred to as *gale commun,
gale ordinaire,* or *gale profonde* in French. Spanish names for this disease
include *roña, sarna común, sarna morena, actinomycosis,* and *sarna de
América.* The Germans call it *gewöhnlicker Kartoffelschorf, Schorf,* or
gewönnlicker Schorf (Miller and Pollard, 1976). Common scab occurs in
Africa, Asia, Australia, Europe, North America, and South America.

A. Importance

Common scab is a serious disease of potato tubers. It produces insignif-
icant or no symptoms above ground. It has little or no effect on total yield,
but it has a tremendous effect on the marketability of the crop. According
to Walker (1969), "Scab remains one of the most important and least satisfactorily controlled of potato diseases."

B. Causal Agent

The causal agent of common scab is the actinomycete *Streptomyces scabies* (Thaxt.) Waksman and Henrici. Thaxter first identified and described
the pathogen in 1890, calling it *Oospora scabies.* The name was changed to
Actinomyces scabies (Thaxt.) Gussow in 1914. Some writers classify it as a
bacterium (Breed *et al.,* 1957; Walker, 1969), while others, including the
United States Department of Agriculture, usually classify it as a fungus
(Anonymous, 1960; Drechsler, 1919; O'Brien and Rich, 1976). It produces
a rudimentary, coiled hyaline mycelium characteristic of fungi. Septations
then form, resulting in one-celled hyaline spores $1-2 \times 0.6-0.7$ μm resembling bacteria. These spores germinate by means of one or two germ tubes.
Some systematists consider the actinomycetes a connecting link between the
bacteria and true fungi, but Drechsler disagrees.

C. Symptomatology

Two types of tuber symptoms are often recognized. One is called shallow, corky, surface, or russet scab; the other is called deep or pitted scab. Shallow scab lesions vary considerably in size. They may coalesce, resulting in a diffuse russet appearance (Fig. 2.4), or they may be slightly raised to slightly sunken. They result from abnormal proliferation of the cells of the periderm. Russet-skinned cultivars tend to be susceptible to this type of scab. Harrison (1962) considers russet and common scab to be two distinct diseases.

Fig. 2.4. Russet Burbank potato tuber infected with common scab caused by *Streptomyces scabies* (top), and Green Mountain tuber infected with pitted scab (bottom).

Deep or pitted scab usually produces distinct brown, corky, sunken lesions of variable size. They are roughly circular (Fig. 2.4), and the majority of them 1–5 mm in diameter. Scab gnats or other insects may invade the lesions thus enlarging them. Thin-skinned red or white cultivars are usually susceptible.

D. Epidemiology

The pathogen survives indefinitely in infested soil, spreading from one location to another primarily by infected seed tubers. It is also disseminated by infested soil which may be transported by wind, water, or mechanically. The organism may also be spread by contaminated manure after passing through the digestive tract of animals (Walker, 1969).

Some research workers have identified pathogenic strains of the scab pathogen (Millard and Burr, 1926). Others believe that more than one species of *Streptomyces* is involved. Harrison (1962) blames an unnamed species of *Streptomyces* as the incitant of russet scab in Minnesota. An acid-tolerant type of scab, found in Maine, has been referred to as "uncommon scab" or "acid scab" (Manzer *et al.* 1977). Bonde and McIntyre (1968) attribute this to a new species of *Streptomyces*. McCrum and Manzer (1967) demonstrated serological differences between species of *Streptomyces*.

Streptomyces scabies usually thrives best in soils with a pH range of 5.5 to 7.5, and rarely has been a problem when soils were maintained at a pH of 5.0 to 5.3. However, in recent years scab has become prevalent and severe in certain soils with a pH below 5.0, giving rise to the names "uncommon scab" and "acid scab" to differentiate it from so-called common scab.

In Washington, land freshly cleared from sagebrush produced scab-free potatoes. However, after continuous cropping to potatoes for 3 years or more, scab became a serious problem. In Rhode Island, however, land with a history of scab produced scab-free potatoes after alfalfa was grown on it for 4 years. Crop sequence, therefore, has a marked effect on the presence or absence of scab (A. E. Rich, unpublished data).

According to Chupp and Sherf (1960), scab occurs on fleshy roots or tubers of beets, cabbage, carrot, eggplant, mangel, onion, parsnip, radish, salsify, spinach, and turnip. The organism can also cause damping-off.

E. Control

Scab-free seed potatoes should be planted in scab-free soil. Usually potato scab can be controlled by maintaining the soil pH between 5.0 and 5.3 (Hooker, 1957). This is relatively easy where the soils are naturally acid. The use of fresh manure, wood ashes, or excessive lime should, however, be avoided. If soils are too acid, dolomitic limestone should be applied following potatoes in the rotation. Gries and Horsfall (1946) concluded that a high calcium–potassium ratio in the soil favors infection. Soluble manganese may also influence scab development (Mortvedt *et al.,* 1961), as does increased nitrogen (Lapwood and Dyson, 1966). Menzies (1957, 1959) discovered a biological factor in some soils which suppressed potato scab.

The use of soy beans (Oswald and Lorenz, 1956; Weinhold *et al.,* 1964a,b) and alfalfa (A. E. Rich, unpublished data) in the rotation reduces the severity of scab. Long rotations with nonsusceptible crops will reduce the inoculum potential in the soil.

The most effective method of scab control is to plant resistant varieties. Most of the russet-skinned cultivars, such as Early Gem, Russet Burbank (Netted Gem), Russet Rural, and Russet Sebago, are partially resistant (Stevenson *et al.,* 1955; Talburt and Smith, 1975; Weber, 1947). Akeley *et al.* (1961), Blodgett and Stevenson (1946), Edmundson *et al.* (1961), Rieman and Young (1955), Sanford, *et al.* (1964), Schaal *et al.* (1949), Wheeler *et al.* (1944), Young *et al.* (1962), and Walker (1941, 1953, 1965) list the following cultivars as resistant: Ackersegen, Antigo, Arnica, Avon, Blanca, Catoosa, Cayuga, Cherokee, Dauerragis Aal, Early Gem, Erdgold, Hindenburg, Huron, Knick, La Rouge, Menominee, Navajo, Norland, Ona, Onaway, Ontario, Osage, Ostragis, Plymouth, Pungo, Red Skin, Reliance, Richter's Jubel, Russet Burbank, Russet Rural, Seneca, Shoshoni, Superior, Tawa, Treff As, and Yampa. Some newer scab-resistant cultivars include: Alamo (Akeley *et al.,* 1968), Chinook (Nonnecke *et al.,* 1966), Hi-Plains and Platte (O'Keefe and Werner, 1965), Nampa and Targhee (Pavek *et al.,* 1973a,b), Nooksack (Hoyman and Holland, 1974), Norchief and Norchip (Johansen *et al.,* 1969a,b), Norgold Russet (Johansen, 1965), Pennchip (Mills, 1964), Pride (Pratt, 1969), Sable (Davies and Young, 1966), Shurchip and Sioux (O'Keefe, 1970a,b), Wauseon (Cunningham *et al.,* 1968), and Wyred (Riedl, 1968). Some cultivars have a higher degree of resistance than others. Individual response will vary with inoculum potential, pH, soil moisture, crop sequence, fertility level and balance, and the species or pathogenic races of *Streptomyces* involved.

Murphy *et al.* (1982) list Belchip, Belleisle, Caribe, Norchip, Rideau, Trent, Wauseon, AF 186-5, B8943-4, C7358-26A, CA02-7, and W564-3A as resistant to common scab. They also classify the following seedling selections as resistant to acid scab: AF92-3, AF186-2, BR7093-23, C7358-14A, C7358-26A, CD106-16, and W564-3A. Thus it appears probable that commercial cultivars which are resistant to acid scab will be available in the near future.

Scab is generally favored by low soil moisture. Thus, irrigation has been suggested as a means of control (Davis *et al.,* 1972; McKee, 1968). Soil treatment with chemicals has also been suggested (Erickson, 1960; Houghland and Cash, 1957; Schultz *et al.,* 1961; Weinhold *et al.,* 1964b). In Idaho, soil treatment with sulfur, gypsum, or pentachloronitrobenzene (PCNB) reduced the incidence and severity of scab (Davis *et al.,* 1974). Uracide was ineffective on Long Island, New York (Cetas and Sawyer, 1962).

VI. PINK EYE

Apparently pink eye is a relatively new disease of potato tubers in North America. It is also referred to as *brown eye* in English or *oeil brun* in French (Conners, 1967). Dale (1912) reported a similar disorder in Great Britain which he called "blindness."

A. Importance

Pink eye is a relatively unimportant disease of potatoes. It is a problem only on a few susceptible cultivars. It usually occurs only on tubers from plants which were infected with Verticillium wilt (Dale, 1912, Folsom *et al.,* 1951; Frank *et al.,* 1973; Hodgson *et al.,* 1973; Robinson *et al.,* 1957).

The Kennebec cultivar is more susceptible to pink eye. Other susceptible cultivars include Katahdin, Russet Burbank (Netted Gem), and Sebago. It occurs in the Maritime Provinces and Quebec in eastern Canada. It is confined mostly to Kennebec in Maine, New Hampshire, and other northeastern states. In Wisconsin, it is called "brown eye," and it is a problem on Kennebec. It occurs less frequently on Cherokee, Chippewa, Irish Cobbler, Katahdin, Sebago, and White Rose. One should be careful not to confuse it with the normally pink eyes which occur on Warba and some other cultivars. Mild pink eye symptoms would be difficult to detect on red-skinned cultivars.

B. Causal Agent

Pink eye appears to be associated with Verticillium wilt. Apparently Verticillium wilt predisposes the tubers to attack by soil bacteria which are normally saprophytic or only weakly pathogenic. Frank *et al.* (1973) demonstrated a relationship between pink eye and Verticillium wilt in Maine, which had been suggested earlier by Friedman and Folsom (1953) and Folsom and Friedman (1959). Cetas (1971), on the other hand, found a lack of correlation between these two diseases on Long Island, New York.

Pseudomonas spp. have been repeatedly isolated from affected tubers in Maine and Wisconsin, but workers disagree on the species identification.

Robinson *et al.* (1957) ruled out *P. lachrymans* and *P. fluorescens,* but stated that it conformed very closely to *P. effusa.* Frank *et al.* (1973) and Hodgson *et al.* (1973) list *Pseudomonas fluorescens* (Flügge) Migula as the causal organism of pink eye. According to Agrios (1969) there are 90 plant pathogenic species of *Pseudomonas,* some of which are very difficult to identify accurately. *Pseudomonas* is a gram-negative, short-rod bacterium with 1–5 polar flagella. It produces a green fluorescent pigment in culture. Garrard (1946) described a storage rot of potatoes in Canada caused by an organism resembling *P. fluorescens* that was favored by warm storage conditions. Frank *et al.* (1973) also found that *Rhizoctonia* can sometimes induce or stimulate development of pink eye in tubers.

C. Symptomatology

Shallow, pink to brown patches are visible around the eyes of affected tubers at harvest time. Discolored patches may also occur between the eyes, especially near the eye end of the tuber (Fig. 2.5). Moderate to severe symptoms can be mistaken for mild symptoms of late blight tuber rot. The affected areas usually dry up under cool, dry storage conditions. Conversely, under warm, humid conditions the symptoms may progress until tubers develop internal reddish-brown to black areas, cavities, and soft rot. This syndrome has been termed high-temperature breakdown (Folsom *et al.,* 1955; Frank *et al.,* 1973; Hodgson *et al.,* 1973; Robinson *et al.,* 1957). Signs consist of short-rod, gram-negative bacteria belonging to the genus *Pseudomonas.*

D. Epidemiology

Verticillium wilt favors the development of pink eye. *Rhizoctonia* may also play a role. Apparently these diseases predispose the tubers to attack by relatively nonpathogenic species of *Pseudomonas* bacteria. High soil temperature and moisture levels may also favor disease development. It is most prevalent on the highly susceptible Kennebec cultivar. High storage temperatures and humidity levels favor symptom development in storage, which may lead to breakdown of the tubers.

E. Control

Control of Verticillium wilt is very important in lessening the chances of pink eye. Highly susceptible cultivars such as Kennebec should be avoided. Affected tubers should be stored at cool temperatures where the humidity

Fig. 2.5.   Potato tubers showing pink eye symptoms associated with *Verticillium* fungi and *Pseudomonas* bacteria.

is relatively low. Treating seed potatoes with benomyl will aid in the control of pink eye.

VII. RING ROT

Ring rot is one of the most contagious and most feared diseases of potatoes, especially among seed potato growers. It is called bacterial ring rot, or simply ring rot in English. At one time it was called bacterial wilt and ring rot, but this tended to be confused with southern bacterial wilt, so the term "wilt" was dropped, even though it is a bacterial wilt disease. The

French names for this disease are *Flétrissure bactérienne* and *bactériose annulaire.* Spanish-speaking people call it *pudrición anular* or *pudrición bacterial anular.* In German it is referred to as *Bacterienringfaüle, bakterielle Ringfaüle,* or *Ringbakteriose* (Miller and Pollard, 1976).

It occurs in Asia, Europe, North America, Central America, and South America. It was first found and described in Germany by Spieckermann in 1913. Apparently it was introduced into Canada from Europe by 1931 and was first found in Maine in 1932. It spread very rapidly to most of other potato-growing states, either in infected seed potatoes or table stock (Walker, 1969).

A. Importance

Ring rot can cause large losses in tonnage as a result of tuber breakdown, but it may cause even greater economic losses due to rejection of entire lots for seed purposes. Most certifying agencies specify a zero tolerance for this disease, thus, one infected plant in a field or one infected tuber in the field, bin, or shipment disqualifies the entire lot for certification. Because the disease is so contagious, rigid rules are essential. In spite of this, the disease still seems to spread.

B. Causal Agent

Ring rot is caused by *Corynebacterium sepedonicum* (Spieck. and Kotth.) Skapt. & Burkh. The bacterium is a nonmotile, gram-positive, short rod, 0.8–1.2 × 0.4–0.6 μm. Agar colonies are white, translucent, glistening, thin, smooth, and small (Walker, 1957; Weber, 1973). Diagnosis of the disease is usually based on a modified gram stain.

C. Symptomatology

Plants infected with ring rot usually fail to show symptoms until late in the growing season, about 80–120 days after planting. It is possible that mildly infected plants never show recognizable symptoms or they may be masked by other disease symptoms such as leafroll (Nelson and Torfason, 1974). Characteristic symptoms are the wilting of leaves on one or more stems in a hill while other stems may appear normal. Affected leaves first become pale green and then chlorotic. This is followed by marginal necrosis. Death of the leaves and branches or the entire plant finally ensues (Fig. 2.6). A milky bacterial exudate can be squeezed from the cut portion of the base of an infected stem (Bonde, 1939a; O'Brien and Rich, 1976; Rich, 1968).

Fig. 2.6. Potato plant infected with bacterial ring rot caused by *Corynebacterium sepedonicum.*

This disease probably went unrecognized for many years because the symptoms often resemble Verticillium wilt, Fusarium wilt, and blackleg. In addition, field inspection of seed potatoes occurred too early in the season to recognize the typically late-appearing symptoms. Symptoms are often masked under cool growing conditions. Sometimes frost kills the plants before symptoms are expressed clearly. Guthrie (1959b) recognized early, dwarfing symptoms of infected plants produced by infected tubers in Idaho. Plants have short internodes and velvety, malformed terminal leaves, often exhibiting necrotic margins. Wilting and early death soon follow as symptoms progress.

Infected tubers develop characteristic symptoms, both externally and internally. External symptoms consist of ragged cracks on the surface of infected tubers (Fig. 2.7). If the stem end of a mildly infected tuber is removed, the vascular ring may exhibit a yellowish discoloration. A yellowish, cheesy ooze may exude from the vascular ring if pressure is applied to the tuber. Severely infected tubers show a darker discoloration of the vascular ring and more advanced stages of rot (Fig. 2.7). Soft rot bacteria often invade infected tubers resulting in their more or less complete breakdown.

Signs of the disease are the very numerous short-rod bacteria which are gram-positive when stained. A modified gram stain (Racicot *et al.,* 1938) is a very useful diagnostic tool to differentiate the causal agent from *Erwinia atroseptica, E. carotovora, Pseudomonas* spp., and other gram-negative bacteria. *Corynebacterium sepedonicum* is the only gram-positive bacterium which causes a disease of potatoes.

The staining procedure used to identify these bacteria is as follows. First, smear a very little ooze from a stem or tuber suspected of ring rot infection on a microscope slide. Then heat the slide very gently to "fix" the bacteria on the slide. Make up the following solutions.

(1)	Crystal violet (or gentian violet)	2.5 gm
	Water	1000 ml
(2)	Sodium bicarbonate	12.5 gm
	Water	1000 ml
(3)	Iodine	20 gm
	Sodium hydroxide (molar solution)	100 ml
	Water	900 ml

Fig. 2.7. Potato tubers infected with bacterial ring rot caused by *Cornybacterium sepedonicum.*

Dissolve the iodine in the sodium hydroxide solution, and dilute with the water.

(4) Ethyl alcohol, 95%	750 ml
Acetone	250 ml
(5) Basic fuchsin, saturated solution in 95% alcohol	100 ml
Water	900 ml

Flood the smear with equal parts of (1) and (2) for about 10 sec, then drain off the excess. Flood with (3) for about 10, and then wash with water. Flood with (4) until no more color comes away, about 5 to 10 sec, then wash with water. Flood with (5) for not over 2 or 3 sec; wash with water and then dry. For immediate examination, smears can be blotted lightly with filter paper or paper towel. A drop of immersion oil can then be placed on the smear, and the slide can be examined under the oil immersion objective of the microscope. Numerous, tiny, short-rod gram-positive (blue) bacteria will indicate a positive reading for ring rot.

The ooze test alone is only about 97% accurate, according to Knorr (1945); Davidson (1946) claims it is unreliable. Stobel and Rai (1968) described a serodiagnostic test for ring rot which they found to be both effective and rapid.

D. Epidemiology

As stated earlier, bacterial ring rot is very contagious. Bacteria are spread by the cutting knife, picker planters, containers, such as baskets, burlap bags, barrels, and storage bins, and equipment such as harvesters and graders (O'Brien and Rich, 1976; Rich, 1968). Duncan and Généreux (1960) also reported their spread by several different groups of insects.

Bacteria overwinter primarily in infected tubers, either during storage or those tubers left in the ground. It is apparent that the disease spreads rapidly from one area to another in infected seed potatoes. For example, shortly after the disease was discovered in Maine it was found within a few years in 36 other states. In spite of a zero tolerance for this disease in seed potatoes, it seems to escape detection occasionally. For example, ring rot was found on a foundation seed potato farm in 1976 which had remained apparently free from this disease for 28 years previously (Eastman, 1976).

Fortunately *C. sepedonicum* does not overwinter in the soil per se. It can, however, overwinter in infected tubers left in the soil, and the disease can be carried along in volunteer plants in succeeding years (Bonde, 1942). It can also survive for extended periods on machinery, containers, etc. There is some danger of spread down the row in irrigation water, or from plant to plant by root contact.

The organism is primarily a wound parasite. It does not appear to enter healthy tissue through lenticels or stomata. Symptoms appear more quickly when young plants are inoculated through wounded roots than when cut seed pieces are inoculated.

Under natural conditions, potatoes are the only hosts. However, tomato, eggplant, and some other Solanaceae have been infected experimentally (Walker, 1969).

E. Control

The only way to control ring rot is to use absolutely disease-free seed and to practice strict sanitation. Even with a zero tolerance for certified seed, the disease is either masked or occasionally escapes the attention of inspectors. Seed potatoes should be handled only in new containers. Seed handlers should be provided with disinfested knives and new gloves (if gloves are worn). Planters and other equipment should be thoroughly disinfested with boiling water and copper sulfate (1 lb in 5 gallons H_2O or 1 kg/40 liters H_2O) or some other effective disinfestant. Copper sulfate may corrode metal equipment. Knives can be dipped in alcohol and flamed or placed in boiling water for several minutes. Ethylene oxide may be used to disinfest storages, bags, etc.

Several investigators have suggested the use of rotary knives which are continually being disinfested by rotation through a disinfestant solution. However, this practice may only mask symptoms further. Planting small, whole tubers is preferable.

If ring rot is found on a seed potato farm, all potatoes should be sold for table stock, all potato-handling equipment and storages should be thoroughly disinfested, and new foundation seed potatoes should be purchased. Many growers have successfully eliminated ring rot from their farms by extremely careful management.

Resistant United States cultivars include Merrimack, Saranac, and Teton, while Russian cultivars reported to be resistant are Rissinovechesky and Minsky (O'Brien and Rich, 1976; Rich, 1968; Riedl, 1946; Walker, 1965, 1969).

VIII. MISCELLANEOUS BACTERIAL DISEASES

Occasional reports of the presence of bacteria in healthy potato tissue have appeared in the literature. Lutman and Wheeler (1948) identified the predominant organism in healthy tubers as *Bacillus megaterium*. Hollis (1949) recovered a mixed, heterogeneous bacterial flora from the vascular tissue of all the parts of potato plants and tubers except the true seeds. Conners (1967) and Jackson and Henry (1946) reported the pathogenicity of *Bacillus polymyxa* (Prag.) Migula on potato tubers in Alberta, Canada.

Folsom *et al.* (1948) described a red xylem disease of potato tubers in Maine. As the name implies, the xylem or vascular ring develops a reddish discoloration. The Katahdin cultivar is most commonly affected. The bacteria are unlike those which cause soft rot or blackleg. The pathogen disappears from affected tubers during extended storage periods.

Malcolmson (1960) described a new disease of potatoes and tomatoes caused by *Bacillus subtilis.* She first observed black spots on the upper surface of affected leaves resembling drops of soil. Veins on the lower leaf surface were black, surrounded by dark green, water-soaked tissue. Axillary buds were dead, which was followed by dieback of affected shoots. On the tubers, many eyes were dead, particularly at the rose (eye) end. Occasionally the adjacent tissue rotted. Apparently the disease is tuber-perpetuated.

Klein *et al.* (1976) described a new disease of potato plants, termed leaflet stunt. It is characterized by deformation and stunting of leaflets. The causal agent is a tiny rod-shaped particle, 1.0–240 × 0.2–0.3 μm. It resembles a mycoplasma but has a double membrane.

Bacillus mesentericus Trev. is recorded as the causal agent of a slimy soft rot. It is usually secondary but is sometimes primary (Anonymous, 1960). *Bacterium polymorphum* (Frankland and Frankland) Migula was reported to cause seed piece rot in Rhode Island (Anonymous, 1960).

Fungus Diseases

I. INTRODUCTION

Fungi are thallophytic plants which lack chlorophyll. Although many fungi are saprophytes, a great many of them are parasitic on higher plants including potatoes. Unlike bacteria, most fungi reproduce by spores. They are also characterized by the presence of vegetative hyphae or mycelium. The Phycomycetes are characterized by the presence of nonseptate or coenocytic mycelium. The Ascomycetes, Basidiomycetes, and Fungi Imperfecti have septate mycelium (Alexopoulos, 1952).

A fungus may produce asexual spores, sexual spores, or both. Ascomycetes produce sexual spores (typically eight), called ascospores, in an ascus or sac. Basidiomycetes produce sexual spores (typically four), called basidiospores, on a basidium or club. In the Fungi Imperfecti or Deuteromycetes, sexual spores are lacking or are unknown (Alexopoulos, 1952).

Some fungi cause leaf diseases, others cause tuber diseases, and many of them produce symptoms on both plants and tubers. Probably the best known fungus disease of potatoes is late blight which was responsible for the Irish Famine in the mid-nineteenth century.

The most important fungus diseases will be discussed in alphabetical order of their common names. The less important fungus diseases will be discussed under a miscellaneous heading at the end of the chapter.

II. BLACK DOT

Black dot is also referred to as anthracnose or root rot in English. French names for this disease are *dartrose* and *anthracnose*. Spanish-speaking people call it *antracnosis* or *enfermedad de punto negro*. The Germans call it *Fusskrankeit, Anthraknose,* or *Blattdürre* (Miller and Pollard, 1976).

A. Importance

Although this disease is not as well known or spectacular as some others, it occurs in Africa, Asia, Australia, Europe, North America, Central America, and South America. Dickson (1926) reported that in one lot 27% of the tubers produced diseased plants. In Tasmania, 50% of tubers from an infected area were unmarketable (Wade, 1949). It is also reported to cause a root rot and fruit rot of tomatoes (Conners, 1967).

B. Causal Agent

American writers usually attribute this disease to *Colletotrichum atramentarium* (Berk. & Br.) Taubenh., a member of the Fungi Imperfecti (Anonymous, 1960; Miller and Pollard, 1976; Weber, 1973). Canadian and English writers, on the other hand, classify the causal agent as *C. coccodes* (Wallr.) Hughes (Conners, 1967; Lapwood and Hide, 1971). According to Weber (1973), "The sclerotia are black, with or without setae, and up to 2 mm in diameter. The conidiophors form a palisade layer, supporting conidia that are hyaline, pink en masse, 1-celled, curved, and 17–22 × 3–8 μ." Wade (1949), in Tasmania, described the sclerotia as 300 to 500 μm in diameter, and the conidia as usually biguttulate but sometimes containing up to five oil drops, 20–56 × 4–12 μm, mostly 25 × 6 μm.

C. Symptomatology

The first field symptoms are a yellowing of the upper leaflets, followed by an upward rolling. Affected leaves gradually droop and die. Plants are stunted and mature early. Aerial tubers may develop. The lower portion of the stem and affected stolons turn reddish-purple, and the cortex breaks down. Tubers are stunted and may exhibit irregular gray patches on the surface of the skin or patches of pseudosclerotia or black dots. Black pseudosclerotia may form on lower stems and stolons also, hence its name. Other signs are described under the causal agent.

Roots turn brown, then black, giving rise to the name root rot, especially on tomatoes. Affected plants may wilt at mid-day and recover at night. Thus they may be confused with Fusarium or Verticillium wilt.

D. Etiology

Colletotrichum atramentarium is usually a weakly parasitic soil fungus which parasitizes tomatoes, potatoes, peppers, and eggplants when grown under certain adverse conditions. It is favored by continuous cropping of the above vegetables which allows a build-up of inoculum in the soil. It may also parasitize cole crops and cucurbits.

The fungus can overwinter in the soil or on infected tubers. It can also be spread from place to place by infested soil or infected seed potatoes. Sclerotia can survive in the soil for 2 years (Chupp and Sherf, 1960; Hodgson *et al.,* 1974). In Tasmania it was usually associated with powdery scab *(Spongospora subterranea)* lesions or more rarely with enlarged lenticels. Some form of injury was necessary for infection (Wade, 1949).

E. Control

Crop rotation with cereals, grasses, or other nonhost plants is highly desirable. Good cultural practices, such as fertilization, irrigation, good soil drainage, and cultivation, will minimize losses from this disease. Disease-free seed potatoes should always be planted. Apparently there are no resistant cultivars which are recommended at present.

III. CANKER OR GANGRENE

The variety of names applied to this disease causes considerable confusion. It is called canker, gangrene, button hole rot, pocket rot, dry rot, and Phoma tuber rot in English (O'Brien and Rich, 1976; Hodgson *et al.,* 1974). French names include *pourriture phomeéne* and *gangrène.* It is called *pudrición de la raíz* or *cancro* in Spanish and *Pustelfaüle* in German (Miller and Pollard, 1976).

A. Importance

Gangrene occurs in Australia, Europe, North America, and South America. It is primarily a storage rot of tubers, but in Europe the causal agent also attacks stems. It is a serious problem in northeastern Scotland, and in Canada it is confined primarily to the Maritime provinces.

B. Causal Agent

The correct nomenclature for the causal agent of this disease is confusing. Miller and Pollard (1976) call it *Phoma solanicola* Prill. & Delacr. Weber (1973) and Hodgson *et al.* (1974) call it *Phoma tuberosa* Melhus, Rosenb., & Schultz. According to Lapwood and Hide (1971) *Phoma exigua* var. *foveata* is the main cause of gangrene in Britain. O'Brien and Rich (1976) list *Phoma exigua* Desm. var. *foveata* (Foister) Boerema as the causal agent and gave the other names as synonyms. *Phoma exigua* var. *exigua* has also been implicated. For further information see Boerema (1967), Lapwood and Hide (1971), or Malcolmson and Gray (1968c).

The fungus is a member of the Fungi Imperfecti. According to Weber (1973), the fungus produces a brown, septate, branched mycelium. Pycnidia are black, submerged to erumpent, scattered, and are up to 160 μm in diameter. Pycnidiospores are hyaline, one-celled, 3.7–6 × 1.8–3.7 μm. Apparently this fungus also causes a similar disease called skin necrosis in Scotland (Foister, 1952).

C. Symptomatology

Tuber symptoms do not show up at harvest time but develop during storage. Small, dark, sunken lesions, resembling thumb marks, develop on the surface of tubers and are usually associated with wounds. They can vary from 6 to 50 mm in size, with well-defined edges. Frequently the skin becomes papery over the sunken lesion and often tears, giving it the appearance of a ragged button hole, thus, the origin of the name "button hole rot." In the early stages, lesions exhibit a light brown watery rot, proceeding to a purplish or black rot flecked with pink. This may be confused with Fusarium rot. Cavities develop as the tissue dries out (Fig. 3.1). Black pycnidia develop both within and on the surface of infected tubers.

In Europe stems are attacked through leaf scars, forming necrotic patches. The older decayed tissue turns white, accompanied by stalk break. Pycnidia on stems are green to yellow at first but later become brown or black with age. Pycnidia on stems are green to yellow at first but later become brown or black with age. Pycnidiospores exude from moistened pycnidia (Lapwood and Hide, 1971; O'Brien and Rich, 1976).

D. Etiology

This disease can be distributed by infected seed potatoes and is primarily a disease of potato tubers during storage. Injuries predispose tubers to infection. Factors which favor healing of wounds, such as a moderate tem-

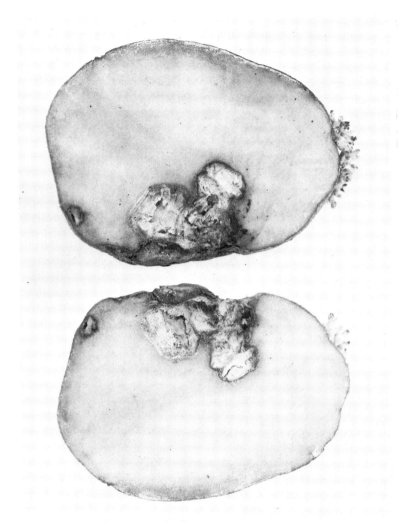

Fig. 3.1. Potato tuber cut to show symptoms of *Phoma* tuber rot. (Photo courtesy of Maine Life Sciences and Agricultural Experiment Station.)

perature and relatively high humidity for a few days following harvest, will hasten wound healing and reduce the incidence and severity of disease. Apparently the fungus can live in the soil for several years. Foister (1952) believes it is ubiquitous. The disease is most serious when tubers have been stored at a cold temperature or moved when cold. Infection occurs before harvesting or during storage from infested soil adhering to the tubers. High soil moisture and poor soil fertility favor disease development. Rapid killing of vines reduces infection (Malcolmson and Gray, 1968a,b).

E. Control

Whenever possible, disease-free seed potatoes should be used. Crop rotation may prove somewhat beneficial. The most important control measure is careful handling at harvest time to prevent bruising, and storage at 15°C for 10 days to promote rapid healing of wounds. Chemical treatment of tubers after harvest with benomyl, thiabendazole (TBZ), or fuberidazole controlled gangrene in storage as effectively as an organo-mercurial dip. A combined application of 1% benomyl and 1% captafol was more effective than either one alone (Copeland and Logan, 1975). Logan *et al.* (1975) recommended an ultra low volume mist of 2% TBZ.

Cariboo is a new cultivar which is reported to be resistant to this disease. Arran Pilot and Golden Wonder are reported to be susceptible to *P. exigua* var. *foveata,* while Kerr's Pink and Doarr Star are more susceptible to *P. exigua* var. *exigua* (Lapwood and Hide, 1971; O'Brien and Rich, 1976). Foister (1952) rates Majestic, Home Guard, and Arran Pilot as very susceptible. The new cultivars Acadia Russet and Caribe are also reported to be resistant to Phoma tuber rot (Murphy *et al.,* 1982).

IV. CHARCOAL ROT

Charcoal rot is a common disease of many vegetables, including potatoes, in most tropical and subtropical countries. It is often called ashy gray stem when it occurs on beans (Chupp and Sherf, 1960).

A. Importance

Charcoal rot is probably present in all tropical and subtropical countries where potatoes are grown. It is serious in the Mediterranean area, in the southern United States and central California, and it occurs as far north as Maryland. The disease is usually or minor importance on potato, but may become severe following a period of unusually warm, wet weather. It can also affect wounded tubers during storage (Chupp and Sherf, 1960; O'Brien and Rich, 1976).

B. Causal Agent

The causal agent of charcoal rot is *Macrophomina phaseolina* (Tassi) G. Goidanich; *M. phaseoli* (Maubl.) Ashby is a synonym (O'Brien and Rich,

1976). *Botryodiplodia solani-tuberosi* Thirum. & O'Brien was also listed as a causal agent of charcoal rot in two recent papers (O'Brien and Thirumalachar, 1977; Thirumalachar and O'Brien, 1977). Weber (1973) considered *Sclerotium bataticola* to be the sterile, sclerotial stage of the fungus, but Chupp and Sherf (1960) state that *S. bataticola* is an entirely different pathogen.

C. Symptomatology

Following seed-piece infection the fungus may grow up the stem to the soil surface and kill the plant. A soft, dark-colored shallow rot develops on the lower stem area, somewhat resembling black leg. Secondary organisms frequently follow primary infection by *M. phaseolina*. The fungus enters tubers through infected stolons. A shrunken, dark-colored area develops at the stem end of infected tubers. The fungus may also invade tubers through eyes and lenticels, thus producing symptoms somewhat resembling late blight.

The fungus produces septate mycelium. Numerous tiny sclerotia form in the stolons. Pycnidia may or may not be formed, depending on the strain of the fungus.

D. Etiology

Under normal conditions, *M. phaseolina* is a weakly parasitic soil fungus. It has a wide host range. It attacks potato plants when the soil is warmer and wetter than optimum for good potato production. Fungal growth and sclerotial development are rare at 10°C or below. The optimum temperature for growth and infection is about 30°C. Poor plant nutrition favors development of this disease. Wounds predispose tubers to infection. The fungus overwinters as sclerotia in soil and plant debris, and can also live from season to season in perennial weeds and other crop plants.

E. Control

Proper nutrition, drainage, and crop rotation should reduce the incidence of this disease. In the sourthern United States it is recommended that growers plant early cultivars, harvest tubers before hot weather, handle them carefully to avoid bruising, and store them in a cool place (O'Brien and Rich, 1976).

A recent study indicated that treatment of seed pieces or whole tubers with a strain of *Bacillus subtilis* Cohn emend. Prazmowski reduced the frequency of charcoal rot at harvest. Apparently *B. subtilis* acts as a bacterial antagonist to the fungus (Thirumalchar and O'Brien, 1977).

V. EARLY BLIGHT

Early blight is an old and well-known disease of potatoes and tomatoes. The name is somewhat misleading because the disease rarely attacks young, vigorously growing plants but often becomes epiphytotic on older more mature plants which have begun to decline in vigor. Other English names for this disease include target spot and leaf spot. French names for this disease are *brûlure alternarienne, alternariose, maladie des taches brunes,* and *taches noires.* Spanish-speaking people call it *mancha negra de la boja, alternariosis, lancha temprana,* and *tizón temprano.* It is referred to as *Dörrfleckenkrankheit, Alternaria-Fäule, Hartfäule,* or *Knollenbefall* in German (Miller and Pollard, 1976).

A. Importance

Early blight occurs in Africa, Asia, Australia, Europe, North America, Central America, and South America (Miller and Pollard, 1976). Probably it occurs almost everywhere that potatoes are grown.

This disease has been underrated in contrast to the more spectacular late blight disease. However, in many areas the average annual loss from this disease exceeds the losses from late blight. The best way to document these losses is to compare yields from some unsprayed plots. Thurston *et al.* (1948) harvested 365 bu from control plots and 586 bu/acre from well sprayed plots in Ohio. Feddersen (1962) increased yields 44% in Australia by spraying with maneb. Control of early blight resulted in 18–39% yield increases in Colorado (Harrison and Venette, 1970). Haware (1968) found that loss in yield rose from 6 to 40% with an increase in disease intensity from 25 to 100%. Early blight can also cause a tuber infection which is often overlooked or misdiagnosed.

B. Causal Agent

The causal agent of early blight is *Alternaria solani* (Ell. & G. Martin) Sor. It is a member of the Fungi Imperfecti. It has also been classified as *Alternaria dauci* f. *solani, A. porri* f. *solani,* and as *Macrosporium solani.* However, *A. solani* is most widely accepted. *Alternaria tenuis* and *A. consortialis* have also been reported on potato in Canada (Conners, 1967).

Cultures of *A. solani* vary widely in pathogenicity. Some are saprophytic while others are highly pathogenic. Henning and Alexander (1959) recognized seven distinct physiologic races of *A. solani* in Ohio.

C. Symptomatology

The characteristic leaf symptoms of early blight are numerous round, oval, or angular dark brown to black, dull necrotic spots. These spots often have concentric rings giving rise to a "bull's eye" or target-board appearance (Fig. 3.2). The spots tend to be vein-limited. Lesions occur first and most abundantly on the lower, senescent leaves which often become yellow. Dark brown to black lesions develop on infected stems (O'Brien and Rich, 1976).

Tuber infection was first identified and described in Maine in 1925 (Folsom and Bonde, 1925). Tuber lesions are usually sunken with raised borders. They are shallow and distinctly set off from healthy tissue by a purplish-brown metallic-hued cork layer (Fig. 3.3) (Miller and Pollard, 1976).

The signs of *A. solani* infection include a light brown, septate mycelium. Condiophores are erect, septate, and measure $5–90 \times 8–9$ μm. Conidia are

Fig. 3.2. Early blight lesions on potato leaf (left) caused by *Alternaria solani*.

Fig. 3.3.   Early blight lesions on potato tubers caused by *Alternaria solani*. (Photo courtesy of Maine Life Sciences and Agriculture Experiment Station.)

obclavate, olive-brown tapering to a long filiform beak, 145–370 × 16–18 μm with 3–14 cross septa and 0–8 longitudinal septa (Western, 1971).

Alternaria consortialis produces light brown spots on potato leaves in Canada. The spots lack the concentric rings typical of early blight caused by *A. solani* (Conners, 1967).

D. Etiology

Alternaria solani overwinters in the field on dead leaves and other plant parts. The disease is favored by short rotations, continuous cropping to potatoes, and by not burning the tops (Manzer and Merriam, 1974). The pathogen can also survive in the soil as a saprophyte. Spores are primarily wind-borne. Infection is favored by warm temperatures and high relative humidity, which is provided by heavy dews, light rain, or irrigation (Easton *et al.,* 1976; Harrison *et al.,* 1965a; Ohms and Fenwick, 1961). High nitrogen and low phosphorus fertilizers resulted in the lowest incidence of *A. solani* infection (Barclay *et al.,* 1973; Soltanpour and Harrison, 1974).

Tiny wounds caused by blowing sand favor disease development, especially if followed by dew, fog, or rain (Rotem, 1959). Insect wounds would probably have a similar effect.

E. Control

The most efficient and practical control measure is to spray with an effective fungicide. Bordeaux mixture and neutral copper fungicides, although effective against late blight, are rather ineffective for control of early blight. Carbamate fungicides, such as maneb, zineb, manganous-zinc carbamates, and captafol have proved effective for control of early blight (see Late Blight, Section VIII). Experiments have shown that three properly timed applications are sufficient. The first application should coincide with the first secondary infection period. Early applications were wasted (Douglas and Groskopp, 1974; Harrison *et al.,* 1965b,c).

Crop rotation and burning of vines where permitted will reduce infection. Tomatoes should not be included in the rotation. Adequate fertilization with high nitrogen and low phosphorus will also reduce disease incidence. Sprinkler irrigation favors disease development and should not be used more than necessary.

Conners (1967) states that early cultivars tend to be highly susceptible. Targhee (Pavek *et al.,* 1973a) is reported to be resistant to *A. solani.* Murphy *et al.* (1982) indicate that Rosa and the following seedlings are resistant to early blight: AF303-5, BR7093-23, CR7358-14A, CR7358-26A, CA02-7, CF7353-1, and CF7523-1. It is likely that several new cultivars will soon be released which are resistant to early blight.

VI. FUSARIUM DRY ROT AND SEED-PIECE DECAY

An attempt is made here to separate tuber rots and decay of potato seed pieces caused by *Fusarium* spp. from Fusarium wilt which is described elsewhere. Fusarium dry rot *(pourriture sèche fusarienne* in French) appears to be the most common name for potato tuber rot caused by *Fusarium* spp. Other names include black rot, side rot, powdery dry rot, fusariosis *(la fusariose),* and stem-end rot *(nécrose fusarienne du talon)* (Blodgett and Rich, 1950; Conners, 1967). The characteristic dry rot of potato seed pieces caused by *Fusarium* spp. is usually called Fusarium seed-piece decay or simply seed-piece decay.

A. Importance

Some Canadian growers lost 50% of their crop, and many growers lost 5–10% of their crops in 1946 and 1947 due to dry rot (Conners, 1967). Numerous reports indicate that Fusarium dry rot is the most serious disease of potatoes during storage. A recent study in the Red River Valley indicated that 46.9% of the tubers were injured during harvesting and bin-filling operations and 19.9% of the tubers developed Fusarium dry rot (Hudson and Orr, 1977).

Seed-piece decay is not a serious problem under normal conditions, but can occasionally become very serious if cut seed potatoes are held too long before planting or are planted in cold, dry soils (Cunningham and Reinking, 1946; Miska and Nelson, 1975. In some cases, stands are so poor that fields have to be harrowed up and replanted with some other crop. In other instances, seed-piece decay may be unjustly blamed for poor stands. A recent study in New Brunswick, Canada, showed that the average crop had 32% misses, but 88% of the misses were due to the seed piece being absent (James *et al.*, 1975).

The nomenclature of *Fusarium* spp. which attack potato tubers and seed pieces is very confusing.

B. Causal Agents

O'Brien and Rich (1976) list *Fusarium sambucinum* Fckl. f. 6 Wr. and *F. caeruleum* (Lib.) Sacc. as causal agents of both Fusarium tuber rot and seed-piece decay. In addition, they list *F. trichotheciodes* Wr. and *F. avenaceum* (Fr.) Sacc. as causal agents of tuber rots. Leach and Webb (1975) list *F. roseum* (Lk.) emend. Snyd. and Hans. *sambucinum* as the causal agent of Fusarium tuber rot. According to Conners (1967), *F. solani* (Mart.) App. and Wr. var. *eumartii* (Carp.) Snyd. & Hansen is the causal agent of stem-end rot. *Fusarium oxysporum*, *F. solani*, and *F. roseum* 'culmorum' and 'sambucinum' have been isolated from stored potatoes in New York and Pennsylvania. Langerfeld (1973) considers *Gibberella cyanogena* to be the perfect stage of *F. sambucinum* f. 6. In Britain, dry rot is caused mostly by *F. caeruleum* and occasionally by *F. avenaceum* (Lapwood and Hide, 1971).

Cunningham and Reinking (1946) identified *F. caeruleum* and *F. sambucinum* f. 6 as the causal agent of seed-piece decay on Long Island, New

York. More recently, Leach and Nielsen (1975) have attributed the decay of seed pieces to *F. solani* and *F. roseum* f. sp. *sambucinum (Gibberella pulicaris)*.

C. Symptomatology

Fusarium spp. usually cause a dry rot of potato tubers in storage, but a moist or wet rot may occur if humidity is high. The surface of infected tubers is wrinkled and may be sunken, and the rotted tissue may turn brown, gray, or black. Cavities frequently develop in affected tubers, which may become more or less filled with yellow, pink or red *Fusarium* molds. After prolonged storage it is common for gray, white, blue, black, purple, or pink spore masses to develop on the surface of infected tubers.

Blodgett and Rich (1950) describe three types of Fusarium tuber rot: black rot, powdery dry rot, and side rot. Black rot [attributed to *F. sambucinum* f. 6. Wr., *F. caeruleum* (Lib.) Sacc., and *F. flocciferum* Cola.] is the most common and severe type of storage rot. Typically, black rot is characterized by moist, firm, discolored tissue beneath an injury, such as a bruise, cut, or where a knob has been broken off. The surface is slightly sunken, but the internal decay may be much more extensive than the external symptoms indicate. At warm temperatures (22°C) the rot is light brown, but under usual storage conditions (5°–15°C) the color is dark brown to black, hence its name. After prolonged storage, a black, blue, or violet fungus growth develops on the surface. Sometimes the entire center of the potato rots, leaving just a "shell," thus giving rise to the term "shell rot." Powdery dry rot (attributed to *F. trichotheciodes* Wr.), is characterized by sunken, shriveled areas filled with powdery, dry, decayed tissue. The rot is mostly gray in color and progresses rapidly until the tuber is a light, punky mummy. The surface and cavities become covered and filled with masses of pink mycelium and spores. Infection usually starts at an injury such as a cut or bruise. Side rot (attributed to *F. eumartii* Carp.) is characterized by a rather firm, dry, granular decay which tends to follow the vascular tissue. It usually starts near the stem end and extends down over the shoulder, thus the name "side rot." It may be synonymous with stem-end rot. Infection usually occurs through the stolon scar, and, eventually, the entire tuber may be mummified. Fruiting bodies of the fungus are less evident than with black rot and powdery dry rot.

Western (1971) describes *F. caeruleum* as follows: "Aerial mycelium white or olive buff-blue; sporodochia brownish white, pink or blue, conidia mostly three-septate 31–40 × 4.5–5.5 μm but occasionally one–two or four–seven septate, sickle shaped usually of even diameter or broader near curved apex; microconidia mostly apedicillate unicellular 16 × 4.7 μm; chlamydospores spheric, terminal or intercalary, smooth, bluish, mostly one-celled, 8-9 μm borne singly or in chains or groups.''

As the name implies, the primary symptom of Fusarium seed-piece decay is a dry rot or decay of infected seed pieces. The decay starts on the cut surface of the seed pieces. Under severe conditions the entire seed piece is decayed. The result is an uneven, poor stand of potato plants in the field. Infected seed pieces may or may not be covered with visible molds of various colors.

D. Etiology

As mentioned earlier, wounds such as cuts, bruises, broken knobs, unhealed stolon scars, etc., predispose harvested tubers to Fusarium dry rot. The various *Fusarium* spp. are present in the soil and invade the tubers through wounds, where they thrive and cause the infected tubers to decay partially or completely. Low initial storage temperatures (2°-5°C) favor initial infection, probably because they do not favor rapid drying and healing of wounds. However, once infection takes place it develops rapidly at higher temperatures (10°-15°C) (Ayers, 1950; Langerfeld, 1973). Jones and Mullen (1974) found that potatoes which were free from potato virus X (PVX) were more susceptible than potatoes infected with a mild strain of PVX.

Fusarium caeruleum can live in the soil for at least 9 years, while *F. avenaceum* is less persistent (Lapwood and Hide, 1971).

Susceptibility to Fusarium dry rot is influenced by the potato cultivar and the fungus species, race, or form involved. For example, Ayers (1961a) lists Sebago, Keswick, Kennebec, Fundy, and Green Mountain as susceptible to *F. sambucinum* f. 6; Kennebec and Green Mountain to be resistant to *F. caeruleum;* but Keswick, Fundy, Irish Cobbler, and Sebago to be susceptible.

Planting unsuberized cut seed pieces in dry soil favors development of seed-piece decay. Storage of cut seed at low temperature with insufficient air retards suberization of the cut surfaces and favors seed-piece decay. Treating cut seed with antibiotics also retards suberization and favors fungus decay.

E. Control

Careful handling to minimize bruising during harvest and storage operations is essential. Equipment should be properly adjusted, padded, and operated. Potatoes should be mature at harvest time. If vines are killed, harvest should be delayed for several weeks for the skin to "set." Potatoes should not be harvested on cold, frosty mornings because they bruise very easily under such conditions. Harvested tubers should be stored at about 10°C and moderate humidity for several days for bruises to heal. The storage temperature should then be reduced to about 5°C.

Cultivars reported to be resistant to one or more types of Fusarium dry rot are Arran Banner, Arran Victory, Kennebec, Green Mountain, Teton, Hudson, Superior, Russet Burbank, Hunter, Epicure, and Belleisle (Ayers, 1961a; Lapwood and Hide, 1971; Leach and Webb, 1975; Munro, 1975). As Doon Star, Arran Pilot, Keswick, and Sebago are very susceptible, they should be avoided.

The surest method of preventing seed-piece decay is to plant small, whole tubers. Small tubers are excellent for seed purposes if they are disease-free. If cut seed is used, it should be well suberized. Fortunately it will usually suberize in the soil if freshly cut seed pieces are planted in a well prepared field which is not too hot, cold, wet, or dry. Cut seed should be stored at about 12°C under fairly humid conditions with good air circulation to promote rapid suberization.

Chemical treatment of either whole seed potatoes or cut seed pieces with an effective fungicide is often recommended. Although in the past, Semesan Bel was recommended by some research workers, it is no longer approved for this purpose in the United States. In addition, the author has observed injury to cut surfaces of tubers following its use. Washing whole tubers with water will reduce the *Fusarium* innoculum, but spraying with a solution of 1200 ppm thiabendazole (TBZ) or benomyl is more effective (Leach, 1975, 1976; Leach and Nielsen, 1975).

Treating cut seed pieces by dusting them with a fungicide, such as captan, benomyl, or metiram, or dipping them in solutions or suspensions of captan, maneb, benomyl, or metiram is good insurance against seed-piece decay. Combinations of fungicides and antibiotics have been tried but are not usually recommended (Ayers, 1974; Bonde and Hyland, 1960; Easton *et al.*, 1970; Miska and Nelson, 1975; Rich *et al.*, 1960).

Walker (1965) lists Irish Cobbler and Hunter as resistant to *F. sambucinum*. He classifies Kennebec as resistant to *F. caeruleum*. Additional cultivars which are resistant to Fusarium tuber rot are Acadia Russet, Belleisle, Shepody, and Tobique (Murphy *et al.,* 1982).

VII. FUSARIUM WILT

This disease is referred to as Fusarium wilt in English to distinguish it from Verticillium wilt and other wilts caused by bacteria such as *Corynebacterium sepedonicum* and *Pseudomonas solanacearum*. French names for it are *flétrissure fusarienne* and *la fusariose.* Spanish-speaking people call it *marchitez por Fusarium, fusariosis,* or *marchitez radical.* German names for it are *Fusarium-Welke der Kartoffel* and *Verwelkingsziekte.*

A. Importance

Apparently Fusarium wilt is distributed in Africa, Asia, Australia, Europe, North America, Central America, and South America. It is difficult to determine how widespread it is in each continent because it has been confused with other wilt diseases, especially Verticillium wilt. Lack of accurate diagnosis makes it extremely difficult to assess correctly the importance of this disease.

B. Causal Agents

It is apparent that considerable confusion exists concerning the correct nomenclature for the causal agents of this disease. Some people prefer Wollenweber and Reinking's (1935) classification, while others prefer Snyder and Hansen's (1940, 1941) or Snyder and Toussoun's (1965) classification. O'Brien and Rich (1976) state that Fusarium wilt is caused by *Fusarium oxysporum* (Schlecht. emend. Snyd. and Hans. and *F. solani* (Mart.) Appel & Wr. var. *eumartii* (Carpenter) Wr. *Fusarium oxysporum* is more widespread but is less pathogenic than *F. solani.* Some workers (Snyder and Toussoun, 1965) believe that *F. oxysporum* is responsible for vacular wilt, while *F. solani* produces a cortical decay. Weber (1973) lists the latter species as the causal agent but further states that many *Fusarium* spp. have been associated with the wilt disease and cause similar symptoms. Hodgson *et al.* (1974) list *F. avenaceum* (Fr.) Sacc. in addition to the above species.

C. Symptomatology

The lower leaves turn yellow and affected plants wilt rapidly. Symptoms resemble those associated with black leg except that the base of the stems do not turn inky black. The symptoms also resemble those of Verticillium wilt. Hodgson *et al.* (1974) state that this wilt can be distinguished from Verticillium wilt by the extensive invasion of areas next to the vascular tissues. A cortical decay of the lower stem, accompanied by necrotic flecks in the pith, may develop. Vascular tissues become brown, especially at the nodes. Vascular ring discoloration of tubers (Fig. 3.4) may be darker and more pronounced than that associated with Verticillium wilt. This disease is favored by hot weather and wet soil. High soil moisture may suppress wilt symptoms, but favors a basal stem rot accompanied by yellowing, roll-

Fig. 3.4. Potato tuber with stem end removed to show vascular ring discoloration caused by *Fusarium* sp. (Photo courtesy of Maine Life Sciences and Agricultural Experiment Station.)

ing, and rosetting of leaves and formation of aerial tubers. This stage can be confused with Rhizoctonia.

Westcott (1971) decribes *Fusarium* as "Mycelium and spores generally bright in color. Macroeonidia fusoid-curved, septate, on branched conidia in slimy masses, sporodochia; smaller micro-conidia with 1 or 2 cells; resting spores, chlamydospores, common. Perfect stage when known usually in Hypocreales, Nectria or Gibberella. Cause of many important rots, wilts, and yellows diseases. Classification difficult, with different systems and synonyms, many forms and races."

The mycelium of *F. solani* var. *eumartii* is hyaline, septate, and branched. The conidia are hyaline, mostly 5-septate, 50–60 × 5–7 μm. They are produced in sporodochia.

D. Epidemiology

Fusarium wilt is generally considered to be a warm weather disease in contrast to Verticillium wilt, which is usually characterized as a cool weather disease. However, the two diseases frequently overlap. It is favored by hot weather and high soil moisture. Irrigation is especially favorable for disease development and dissemination of the pathogens. The fungi which cause Fusarium wilt are common soil pathogens and have a wide host range. They can be distributed in infected seed potatoes or in infested soil adhering to potato tubers. Monoculture and the use of susceptible cultivars favor disease development.

E. Control

Fusarium wilt, like Verticillium wilt, is difficult to control, although control measures for the two diseases are similar. Crop rotation is beneficial. Disease-free seed potatoes should be planted. Seed treatment with an effective fungicide will decontaminate infested seed and may reduce seed-piece decay (Fig. 3.5) (see next section). Potato fields should not be overirrigated.

The use of resistant cultivars will reduce losses from Fusarium wilt. Goss and Jensen (1942) rate Pontiac, Katahdin, Sebago, Arnica, Hindenburg, and Richter's Jubel as resistant, based on tuber symptoms. *Fusarium solani* var. *eumartii* was more important than *F. oxysporum* in their trials. Muncie (1949) rated Teton tubers as resistant to *F. solani* var. *eumartii*.

VIII. LATE BLIGHT

This disease is commonly called late blight to distinguish it from early blight, although both names are often misnomers. Other English names for the disease include potato blight and downy mildew. Maine growers also used to refer to it as "rust." The French name for it is *mildiou*. Spanish-speaking people call it *hielo, pudrición de la raíz, tizón tardio, fitoftora,* or *lancha tardia*. It is called *Kraut-und Knollenfäule, Braunfäule,* or *Kartoffelkrautfäule* in German (Miller and Pollard, 1976).

A. Importance

Late blight is one of the oldest, most publicized, and most serious diseases of the potato. It was first reported from Europe and the United States about 1830. It became increasingly worse in western Europe until 1845 when it was responsible for the Irish Famine. The disease was so serious in Ireland, where potatoes constituted the main diet, that thousands of people died of starvation. Thousands of others emigrated to the United States, Canada, and other countries (Large, 1940).

Rumors have persisted that eating potatoes infected with late blight can

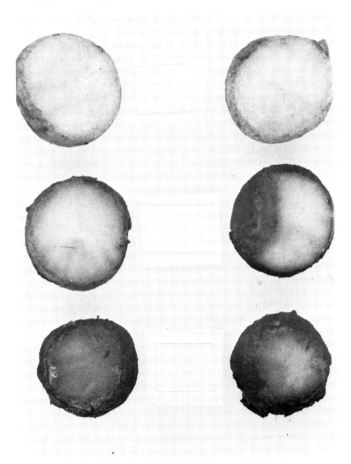

Fig. 3.5. Potato seed pieces showing seed-piece decay, caused by *Fusarium* sp. The two
seed pieces at top of photo were treated with ziram before planting. This treatment effectively
controlled seed-piece decay.

incite spina bifida and anencephaly in humans. However, animal feeding
tests produced negative results (Keeler *et al.,* 1975).

Late blight continued to be an extremely serious disease of potatoes until
the accidental discovery of Bordeaux mixture about 40 years later. It was
used widely for many years until more effective and less phytotoxic fungicides were developed (Large, 1940).

Late blight occurs in Africa, Asia, Australia, Europe, North America,
Central America, and South America. It is most serious in cool, humid
climates such as occur in western Europe, the eastern United States, and

Canada (Miller and Pollard, 1946; Walker, 1969). It became a very serious disease of tomatoes in the United States in 1946, during an unusually cool wet growing season.

B. Causal Agent

The causal organism of potato late blight is *Phytophthora infestans* (Mont.) D By. It is a member of the Phycomycetes. Older names include *Botrytis infestans* Mont. and *Peronospora infestans* (Mont.) D By (Walker, 1969). It is closely related to the downy mildews.

During the decade from 1840 to 1850, considerable controversy raged concerning the cause of late blight. Most people blamed the weather. Even when a fungus was found on the dead and dying foliage it was considered to be saprophytic. The Rev. M. J. Berkeley was the principal proponent of the fungus hypothesis. Dr. Montagne first described the fungus in 1845, calling it *Botrytis infestans*. In 1846, De Bary renamed it *Phytophthora infestans* (Mont.) D By. This was the beginning of the end of the spontaneous generation theory of the disease and the birth of the germ theory of disease, a giant step in the biological sciences (Large, 1940).

C. Symptomatology

This disease affects leaves, stems, and tubers. Water-soaked spots or lesions first appear on the leaves during cool, wet weather. The spot appears light green at first and then turns brown. Lesions may also have a yellowish-green margin or halo. A white fungus ring of conidiophores and conidia develops on the under surface of the leaves near the margin of the lesion if the weather remains wet or humid (Fig. 3.6). This helps distinguish it from Botrytis blight which has dark brown or dark gray conidiophores and conidia. It does not have concentric rings, nor is it restricted by major veins which is typical of early blight lesions. The blight lesions spread rapidly under favorable weather conditions, attacking petioles and stems. Under very severe conditions entire plants or even entire fields may be killed (Figs. 3.7 and 3.8). If the disease is widespread, it may produce a characteristic fetid or decaying odor. Old time potato growers said that they could smell "rust." The spores wash down into the soil and attack the tubers. Tubers also may become inoculated during the harvesting process if they come in contact with infected plants. Initial tuber infection results in a shallow, reddish-brown dry rot (Figs. 3.9 and 3.10). Under cool, dry storage conditions the infected tissue will dry up and the infection will progress very slowly,

Fig. 3.6. Potato leaf showing water-soaked lesions and white fungus mold of *Phytophthora infestans,* causal agent of potato late blight.

if at all. If infected potatoes are stored under warm, humid conditions the rot will continue to develop. Frequently, secondary rots invade the blighted tissue and completely rot the tuber.

Signs of disease consist of a nonseptate mycelium in infected plant parts or tubers, sporangiophores or conidiophores and sporangia, or conidia protruding from stomata at the margins of necrotic leaf spots or from lenticels on tubers. Zoospores may or may not develop, depending on temperature and humidity. Oospores rarely occur in nature, but have been found rather commonly in Mexico in recent years (Galindo, 1965a,b; Gallegly and Gal-

Fig. 3.7. Potato field heavily infected with late blight, caused by *Phytophthora infestans.*

Fig. 3.8. Potato plants killed by late blight in unsprayed potato plots.

Fig. 3.9. Potato tuber and tomato fruit infected with late blight.

indo, 1957; Smott *et al.,* 1957). Walker (1969) describes the characteristics of *P. infestans* as follows:

The conidiophores arise from the leaf surface through stomata and from tuber substrate through lenticels; they are hyaline, branched, and indeterminate. The thin-walled, oval, hyaline conidium (21 to 38 by 12 to 23 μ) with an apical papilla is borne at the tip of a branch, and as it approaches maturity, the branch tip swells slightly, proliferates, and turns the attached conidium to the side as elongation of the conidiophore proceeds. A fruiting hypha, then is characterized by periodic swellings indicating the points at which sporulation has taken place. This type of fruiting branch is characteristic of the genus *Phytophthora* and serves to distinguish it from the most closely related genus, *Pythium.* The conidium may germinate directly by means of a germ tube, which may in turn give rise to a terminal secondary conidium, but most commonly the contents of the conidium cleave to form about eight baciliate zoospores which emerge in a group. The flagella . . . develop paddlelike structures at their tips. One flagellum is ciliated while the other is not. After a few minutes of motility the zoospores lose their flagella, come to rest, and germinate by means of germ tubes. The latter occasionally penetrate through the stomata, but more often an appressorium is formed from which a penetration hypha enters directly through the cuticle. The mycelium in the tissue is coenocytic, intracellular, and intercellular. Rudimentary haustoria are found in foliage cells, but in potato tuber cells they are more elaborate, being club-shaped, hooked, or spirally twisted.

Fig. 3.10. Potato tuber cut to show internal symptoms of late blight tuber rot. (Photo courtesy of Maine Life Sciences and Agricultural Experiment Station.)

Western (1971) gives the following description of *P. infestans.:*

Sporangia borne on sympodially branched sporangiophores, found in annuli surrounding necrotic lesions on leaves, mainly on the undersurface; hyphae intercellular with haustoria, varying from short and spheric to long and straight or hooked, entering cells; oogonia, 31–43 µm in diameter, amphigynous antheridia may be found in culture, but thick walled pale yellow-dark brown oospores only form abundantly from mating strains A1 and A2.

D. Etiology

Phytophthora infestans usually overwinters in infected potato tubers in storage. It is distributed from one place to another in slightly infected seed or table stock potatoes. If infected seed pieces are planted or if infected potatoes are discarded or dumped, they will grow and start new infections. Potato dump piles are an important source of spring infections in Maine (Bonde and Schultz, 1943) and probably elsewhere. Infected tubers which survive in the field may also be a source of spring infection.

Late blight is favored by cool, wet weather. The fungus produces an abundance of conidia or sporangiospores under moist conditions. If the temperature is above 55°F (13°C) or if the relative humidity is not quite 100%, these spores are wind-borne and germinate directly. The disease will spread from one field to another in this manner. However, if the temperature is 55°F (13°C) and the humidity is 100%, each sporangium will produce 8–12 zoospores or swarmspores. These spores are disseminated by splashing rain drops and are responsible for very rapid development of the disease in a field which has become initially infected by germinating conidia.

Conidia and zoospores germinate rapidly when the temperature rises above 55°F, especially in the 60°–75°F (16°–24°C) range. Thus, the ideal conditions for an epiphytotic (severe disease outbreak) are night temperatures of about 12°C accompanied by heavy dew or rain, followed by daytime temperatures of 16°–24°C accompanied by high humidity produced by rain, fog, or lingering dew (Walker, 1969).

Many physiologic or pathogenic races of *P. infestans* occur in nature. It is believed that at least 10 dominant genes for resistance, commonly designated as R_1, R_2, etc., occur in *Solanum,* mostly in *S. demissum.* Potato races of *P. infestans* are designated as 1; 2; 3; 4; 1,2; 1,2,3; etc., depending on the *R* genotypes which they infect. At least 15 potato races have been identified but possibly 2^{10} races could occur (Black, 1957, 1960; Black and Malcolmson, 1965a,b; Black *et al.,* 1953; Western, 1971).

Apparently most new races arise by mutation. Some may arise by recombination or parasexuality (Leach and Rich, 1969), while others may arise from sexual reproduction of *P. infestans* where compatible mating types occur such as in Mexico.

E. Control

Tubers infected with late blight should never be planted for seed. Potato dump piles which often serve as a source of primary inoculum should be destroyed or eliminated. If they are near potato fields they should be treated with an herbicide to prevent growth of volunteer potato plants from infected tubers (Folsom *et al.,* 1955).

Spraying or dusting with an effective fungicide is standard procedure in cool, moist climates which favor late blight development. For many years Bordeaux mixture was the standard fungicide for control of late blight. It is a mixture of copper sulfate, hydrated lime, and water. Although Bordeaux mixture effectively controls *P. infestans,* it is phytotoxic to potatoes. "Neutral" or fixed copper fungicides partially replaced Bordeaux mixture because they were less phytotoxic, easier to prepare, could be used as a dust formulation, and were fairly effective against late blight. However, under conditions favorable for an epiphytotic, Bordeaux mixture has proved more effective than neutral copper dusts (A. E. Rich, unpublished data).

Since World War II, the copper fungicides have been largely replaced by organic fungicides which are effective against both early and late blight and most of them are less phytotoxic to potatoes than are the copper fungicides. Dichlone, usually sold as Phygon, proved to be effective for control of late blight, but caused a reduction in yield when used at rates as low as 1 lb/ acre (Bilbruck and Rich, 1961).

Several carbamate fungicides have proved to be very effective in control of late as well as early blight. Maneb [ethylenebis(dithiocarbamate) manganese] and zineb [ethylene bis(dithiocarbamate) zinc] are two of the most popular and effective carbamate fungicides. Maneb is formulated as a wettable powder, and is commonly marketed under the trade names Dithane M-22 and Manzate in the United States. It is usually applied at the rate of 1½–2 lbs of active ingredient per acre in 100–150 gallons of water (1.70–2.25 kg/ha in 935–1400 liters of water) (Brandes *et al.,* 1959a; Callbeck, 1960; Fink, 1957; Heuberger *et al.,* 1947; Peterson, 1957).

Zineb spray is usually prepared from nabam and zinc sulfate in the spray tank because it is more economical than zineb in wettable powder form. The sodium ions in nabam [ethylenebis(dithiocarbamate) disodium salt] are replaced by zinc ions for zinc sulfate to form zineb. Growers should follow directions on the package for proportions of nabam and zinc sulfate. Because nabam is a liquid it cannot be used in dust formulations. Therefore, zineb is commonly used in preparing dusts. Zineb can be purchased in the United States under such trade names as Dithane Z-78 and Parzate. In foreign countries other trade names are used. In recent years a combination of manganese and zinc ion carbamates have been formulated and marketed under such trade names as Dithane M-45 and Manzate 200. Metiram is another complex dithiocarbamate fungicide which has proved effective for control of late blight. It is available in the United States under the trade name Polyram.

Captafol appears to be one of the most effective of the newer fungicides for control of late blight of potato. It also controls early blight and Botrytis blight. Maneb appears to be somewhat ineffective in its control of Botrytis blight (A. E. Rich, unpublished data). Captafol is marketed as Difolatan in the United States and Canada. It is formulated as a suspension or flowable, rather viscous liquid containing 4 lbs of captafol per gallon. The recommended rate of application is 1–1½ quarts per acre (2.8–4.2 liters/

ha). Ridomil, a new systemic fungicide, also shows promise (Easton and Nagle, 1981).

Other fungicides sometimes recommended for use, especially in the United States and Canada, include chlorothalonil (Bravo) and fentin hydroxide (Du-Ter). Growers should follow the recommendations on the package as to concentrations, rates to use, compatibilities, precautions, etc.

Fungicides can be applied as sprays, dusts, or through the irrigation system. Potter (1981) refers to this method of application as "fungigation." Personal experience has proven that sprays are more effective than dusts under epiphytotic conditions. Dusts should be applied in early morning or in the evening when foliage is damp with dew and air currents are at a minimum. Dusting in the evening is preferred. Dusting causes less soil compaction, requires less time, and does not require a water supply.

Sprays are often applied on a 7 to 10 day schedule. They may be used diluted (about 150 gal/acre or 1400 liters/ha) or concentrated, and are usually referred to as low-volume sprays. The same amount of active ingredient is applied to an area but less water is used as a diluent or carrier. Cannon and Callbeck (1965) obtained better control of late blight with high-volume sprays. Potter and Hooker (1969) reported equally good control with low-volume sprays.

Highly concentrated or ultra low-volume sprays are applied by airplane or helicopter. Usually 3–5 gal per acre (28–47 liters/ha) are applied by aircraft. Aircraft spraying has several advantages: it is fast, plants are not damaged, and soils are not compacted. Fungus spores and virus inoculum are not spread from plant to plant by tractor and sprayer wheels. Hodgson *et al.* (1974) recommend aircraft spraying for fields larger than 10 acres (4 ha). Helicopters should be used where there are trees, telephone wires, buildings, etc., surrounding the fields because they are more maneuverable than fixed-wing planes.

Numerous methods of forecasting late blight infection periods and epiphytotics have been devised by scientists in an attempt to reduce the number of fungicide applications by more accurate timing. They are based on temperature, relative humidity, rainfall, or a combination of these factors, and the probability that late blight will develop under the prevailing environmental conditions if inoculum is present. Spore traps are sometimes used in conjunction with weather data. The same criteria cannot be used in all climates to predict infection periods but must be adapted to individual climatic conditions (Beaumont, 1947; Cox and Large, 1960; Fry, 1977; Hirst and Stedman, 1960; Hyre, 1955; Hyre and Bonde, 1955; Hyre *et al.,* 1960; Krause *et al.,* 1975; Wallin, 1962; Weingartner, 1977).

In Holland, England, and Ireland the weather seldom is too warm for late blight development. The four criteria used to predict late blight are (1) night temperature below the dew point for at least 4 hr, (2) night temperature 10°C or above, (3) mean cloudiness not below 0.8 on the following day, at least 0.1 mm of rain the following day. In some parts of the United States the temperature is often too warm for late blight development in July and August. Therefore one criterion for an epiphytotic is that the temperature remain cool (below 21°C) for an extended period of time.

Potato vines should be killed with chemicals, by burning, or by frost after cessation of spraying and prior to harvest to reduce the chances of tuber infection before or during harvesting operations (Bonde and Schultz, 1945). Western (1971) recommends sulfuric acid, diquat, or dinoseb for vine killing and weed control. Sodium arsenite is effective, but its use may not be permitted in some countries due to its poisonous properties. Chemicals which kill rapidly tend to cause an undesirable discoloration of the vascular ring of potato tubers. Copper sulfate effectively controls late blight, kills potato vines and some weeds, and causes a minimum amount of vascular ring discoloration (Rich, 1951).

The development of resistant cultivars or varieties has played an important role in the control of late blight and will continue to do so in the future. However, many problems have developed along the way. Resistance to late blight is governed by numerous genes and gene combinations. A cultivar may be resistant to one or more races of *Phytophthora infestans* and be susceptible to others. The fungus seems to be quite plastic or mutable, and new races may develop quite readily. Foliage may be resistant and tubers may be susceptible, or vice versa. Some genes for resistance occur in other species of *Solanum,* such as *S. demissum,* thus complicating the breeding programs. Some hybrids mature too late or possess other undesirable horticultural traits (Western, 1971). One of the biggest problems is that cultivars which are resistant to late blight may be highly susceptible to other diseases. The Kennebec's susceptibility to Verticillium wilt is a case in point. Nevertheless, many late blight-resistant cultivars have been developed and have met with varying degrees of success. Most of them are not immune but possess moderate to high resistance to one or more races of the fungus.

Sebago is one of the oldest American cultivars which possesses appreciable resistance to *Phytophthora infestans* (Stevenson and Clark, 1938). It is a leading cultivar in the seed potato industry of Prince Edward Island, Canada, and is commonly grown in Florida.

Numerous names of cultivars have been added in recent years to the list of American and Canadian cultivars which are resistant to this pathogen. They include: Alamo, Kennebec, Merrimack, Ona and Saco (Akeley *et al.,* 1948, 1955a,b, 1962b, 1968); Avon, Keswick, Wauseon, Raritan (Munro, 1975); Belleisle (Young and Davies, 1975); Bison (Johansen *et al.,* 1977); Boone (Haynes *et al.,* 1956); Catoosa (Dykstra *et al.,* 1961); Cherokee (Peterson *et al.,* 1954); Chieftain (Weigle *et al.,* 1968); Delus and Plymouth (Stevenson *et al.,* 1954, 1956); Fundy (Young *et al.,* 1960); Nooksack (Hoyman and Holland, 1974); Onaway (Wheeler and Akeley, 1961); Pennchip (Mills, 1964); Pungo (Parker *et al.,* 1954); Reliance (Hoyman *et al.,* 1963); Tawa (Peterson and Hooker, 1958); and York (Johnston *et al.,* 1970).

Anita, Bertita, Conchita, Dorita, Elenita, Erendira, Florita, Gabriela, and Greta are blight-resistant cultivars developed in Mexico during the past two or three decades (Cervantes, 1965; Niederhauser *et al.,* 1959; Niederhauser and Cervantes, 1959).

Some other older cultivars which possess some blight resistance are Russet Sebago (a mutation from Sebago), Ashworth, Calrose, Canso, Cayuga, Chenango, Cortland, Empire, Essex, Fillmore, Glenmeer, Harford, Madison, Menominee, Placid, President, Saranac, Snowdrift, and Virgil (Rich, 1949, 1977; Stevenson and Livermore, 1949; Walker, 1952, 1953). Most of these cultivars are no longer grown. Keswick and Canso are two of the oldest Canadian cultivars which are resistant to late blight.

Some resistant European cultivars include Craig's Beauty, Craig's Snowwhite, and Pentland Ace from Scotland; Epoka and Wulkan from Poland; Aal, Alpha, Ackersegen, Armica, Aquila, Dauerragis, Erdgold, Erica, Falke, Hindenburg, Noordeling, Ostragis, Panther, Pommernbote, Richter's Jubel, Treph As, Urtica, and Voran from Germany; and Dyelskovsky, Kameraz No. 1, Ketskosellsky, Kransnonfimsky, Pushkinsky, Uktussky, and Vralsky from U.S.S.R. (Walker, 1941, 1953, 1965; Young and Young, 1958).

Murphy *et al.* (1982) classify Atlantic, Belchip, Campbell 11, Campbell 13, and Penn 71 as resistant to late blight in the northeastern United States. They also list the following seedling selections as resistant: AF92-3, AF201-25, B6043-WV6, BR5991-WV16, and W564-3A.

Fry (1977) recommends integrated control of late blight by combining polygenic resistance and timing fungicides according to the "Blitecast" prediction of infection periods. He suggests reducing the concentration of fungicide when cultivars with polygenic resistance are grown. Timing of applications according to "Blitecast" predictions was more effective than applying them after each 1.27 cm (0.5 in.) of rain.

IX. LEAK OR WATER ROT OF POTATO TUBERS

Leak is a common disease of potato tubers. It ocurs in Africa, Asia, Australia, New Zealand, Europe, and North America. Other English names include water rot, watery wound rot, and shell rot. The French refer to it as *pourriture aqueuse* or *pourriture pythienne du planton.* Spanish-speaking people call it *pudrición de agua* or *goteo* (Conners, 1967; Miller and Pollard, 1976).

A. Importance

Leak is usually of minor importance unless the soil is wet, potatoes are dug when it is warm, the skin is broken, or tubers are stored where it is warm. In England and Canada it is reported to cause seed-piece decay of noncallused cut sets after planting (Conners 1967; Western, 1971).

B. Causal Agent

The causal agent is a species of *Pythium,* but there is disagreement on the species name. Miller and Pollard (1976) call it *P. vexans* DeBary, while Conners (1967) calls it *Pythium* spp. mainly *P. ultimum* Trow. Western (1971) calls it *P. ultimum* Trow and Walker (1952) refers to it as *P. debaryanum* Hesse. Western (1971) describes *P. ultimum* as follows:

> No reproductive cells seen on potato but in culture; sporangia chiefly terminal and spheric, 12–28 μm in diameter or occasionally intercalary and barrel shaped, 14–17 μm to 23 by 28 μm germinating only by germ tubes; oogonia (averaging 20.6 μm) spheric, terminal, rarely intercalary; antheridium typically sessile arising immediately below oogonium, sharply incurved; oospores (averaging 16.3 μm in diameter) smooth, yellow with thick wall and single globule.

C. Symptomatology

As the name implies, affected tubers develop a watery soft rot and moisture oozes or "leaks" from them (Fig. 3.11). They may or may not show external symptoms. Red-skinned cultivars show a metallic gray discoloration of the skin, while white-skinned cultivars develop a brown discoloration. The flesh of infected tubers is granular, very watery, and the color varies from light yellow to shades of brown to black. The decayed area is usually delineated by a dark brown to black line (Fig. 3.12). Rotted tissue may become pulpy and develop cavities. Often only a shell of sound tissue remains.

Infected seed pieces or sets show similar symptoms. Such seed pieces will produce very weak plants or even "blanks" or "skips." Otherwise there are no other above ground symptoms.

Signs consist of nonseptate mycelium and oospores. However, these signs are not usually apparent (see Causal Agent, Section IX,B).

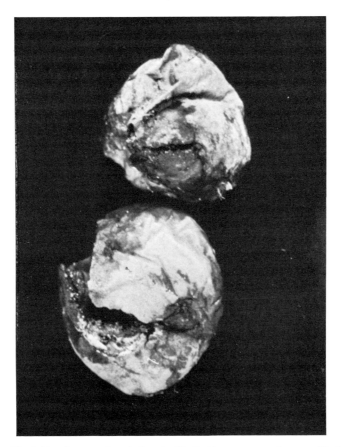

Fig. 3.11. Potato tubers infected with *Pythium* sp., causal agent of leak.

D. Etiology

Pythium is a very common soil fungus. It is responsible for damping-off of seedlings all over the world. The fungus is favored by warm, moist conditions. It usually invades tubers through wounds at harvest time. Tubers are cut, bruised, pierced, or injured in some other manner. Infested soil adheres to the freshly wounded tissue of potato tubers, and the organism grows and multiplies rapidly under warm, moist conditions.

The fungus can survive unfavorable conditions such as dry or cold soils in the form of thick-walled oospores. These spores will germinate to produce nonseptate or coenocytic mycelium when the soil is warm and moist. Potato tissue is an ideal medium for rapid growth of the fungus.

Fig. 3.12. Cross section of potato tuber exhibiting symptoms of leak, caused by *Pythium*. (Photo courtesy of Maine Life Sciences and Agricultural Experiment Station.)

E. Control

The most obvious control measure is to avoid wounding potato tubers during the harvesting process. Potatoes should be mature and machinery should be padded and operated properly. Growers should avoid poorly drained, wet soils. Potatoes should be held in a relatively cool (12–18°C) storage until wounds are healed. They should not be sold or shipped immediately after harvest (Blodgett and Rich, 1950).

Seed pieces or sets should be well suberized before planting. Treating cut seed with an effective fungicide should minimize losses from seed-piece decay (Newton and Lines, 1947). Assisted feed planters which pierce each seed piece make wounds which could serve as a source of entry for the pathogen. Early planting before soils become warm could minimize Pythium seed-piece decay.

X. PINK ROT

Pink rot is a fairly common but usually not too serious disease of potatoes. It occurs in Asia, Australia, Europe, and North America (Miller and Pollard, 1976). A report (Vargos and Nielsen, 1972) indicates that it is present in South America also. English-speaking people usually call it pink rot, but sometimes it is referred to as watery rot or wilt. French names include *pourriture rose* and *pourriture humide.* It is called *podredumbre rosada, pudrición de la raiz, marchitez,* or *pudrición acuosa* in Spanish. German names for it are *Rotfäule* and *Rosafäule* (Miller and Pollard, 1976).

A. Importance

Pink rot is primarily a disease of potato tubers, but occasionally lower stems are infected, followed by wilting of affected plants. It is fairly widespread but not too serious in Canada. It is more common in wet areas of fields or in fields that are irrigated late in the season (Hodgson *et al.* 1974). It occurs widely but sporadically in Great Britain and can be serious in West Eire (Western, 1971).

B. Causal Agents

The causal agent of pink rot is usually considered to be *Phytophthora erythroseptica* Pethybr. (Goss, 1949; Hodgson *et al.,* 1974; O'Brien and Rich, 1976). Western (1971) claims that occasionally it can also be caused by *P. megasperma* Drechs. He describes *P. erythroseptica* as follows:

Sporangiophores 1–2 μm in diameter widening slightly beneath sporangium, generally branching sympodially from immediately below sporangium; sporangia vary in shape, ellipsoid or obpyriform often constricted distal to the middle, nonpapillate 43 $\times$ 26 (maximum 69 $\times$ 47) μm germinate in water while still attached either directly by germ tube or indirectly by zoospores in an extruded vesicle; antheridia amphigynous; oogonia 30–35 (maximum 46) μm, wall smooth; oospores spheric nearly filling oogonium.

A recent report from Ohio (Rowe and Schmitthenner, 1977) indicates that pink rot can be caused by *P. erythroseptica* or *P. cryptogea.* Both fungi were isolated from tubers infected with pink rot, from Ohio soils, and each was independently capable of causing the disease. Other species of *Phytophthora* may also be involved in inciting pink rot (Hooker, 1981; Vargos and Nielsen, 1972).

C. Symptomatology

Pink rot derives its name from the fact that flesh from infected tubers is dirty white when freshly cut, but it turns pink very rapidly upon exposure to air. It becomes a deep salmon pink within one-half hour and changes to brown, then black, after several hours. Uncut tubers are dull brown with dark brown eyes and lenticels (Blodgett and Rich, 1950)

Pink rot is primarily a disease of potato tubers. However, plants may become wilted late in the season, and aerial tubers may form on the stems.

Infected roots and stems turn brown or black, resembling blackleg (Goss, 1949; Hooker, 1981; O'Brien and Rich, 1976).

The signs are described under Causal Agents. Goss (1949) and Rowe and Nielsen (in Hooker, 1981) described the symptoms, signs, and causal agents as they occur in Nebraska and Ohio, respectively.

D. Etiology

Pink rot is favored by warm, wet summers and excessive irrigation. Oospores are formed in stems, roots, stolons, and sometimes in tubers. They are released into the soil when affected tissues disintegrate, and survive from one to several winters. They are the primary source of inoculum for subsequent crops. They germinate to produce zoospores which swim about in the soil moisture and cause infection. The fungus is favored by high soil moisture and a temperature of about 23°C.

E. Control

Only healthy seed potatoes should be planted on land where potatoes have not been grown recently. Wet, poorly drained soils and excessive irrigation should be avoided. Tubers should be handled very carefully during the harvesting process so as to prevent wounding. Only healthy tubers should be stored under relatively cool, dry conditions.

XI. POWDERY SCAB

Powdery scab was first discovered in Germany in 1841 and was found in eastern Canada in 1913. Probably it originated in South America (Walker, 1969).

English-speaking people usually refer to this disease as powdery scab. French names for the disease are *gale poudreuse* and *gale spongieuse.* The Germans call it *Raüde, Pulverschof der Kartoffel, Geschwulste,* or *Schwammschorf.* Spanish names include *roña, sarna polvorienta, roña corchosa, sarna polvosa,* and *roña polvorienta* (Miller and Pollard, 1976).

A. Importance

Powdery scab is a minor disease of potatoes in the United States, but is rather serious in some other countries. Walker (1969) considers it to be the second most important disease incited by a member of the Plasmodiophorales. It is primarily a disease of cool climates (Hines, 1976). It has been reported from Africa, Asia, Australia, Europe, North America, and South America (Miller and Pollard, 1976; Walker, 1969).

B. Causal Agent

Powdery scab is caused by *Spongospora subterranea* (Wallr.) Lagerheim., a member of the Plasmodiophorales. Older names for this fungus include *Erysibe subterranea* Wallr. and *Spongospora solani* Brunchorst (Walker, 1969).

C. Symptomatology

This disease has no foliar or above-ground symptoms. Tuber symptoms may be confused with those of common scab caused by *Streptomyces scabies*. Powdery scab lesions (Fig. 3.13) are smaller and more nearly round than common scab lesions. They are closed or covered by a membrane in the early stages of the disease. Mature lesions are roundish, raised, open pustules and are filled with a brown, powdery mass of spores, surrounded by tattered remnants of the epidermis. Small galls (Fig. 3.14) may develop on the roots of affected plants (O'Brien and Rich, 1976).

Infected tubers may develop dry rot during storage. They are also predisposed to late blight tuber rot (Bonde, 1955). *Spongospora subterranea*

Fig. 3.13. Potato tubers exhibiting symptoms of powdery scab, caused by *Spongospora subterranea*. (Photo courtesy of Maine Life Sciences and Agricultural Experiment Station.)

Fig. 3.14. Powdery scab lesions, caused by *Spongospora subterranea,* on stems and roots of potato plants. (Photo courtesy of Maine Life Sciences and Agricultural Experiment Station.)

may also be a vector of a potato virus disease known as mop top (O'Brien and Rich, 1976).

D. Etiology

Although *S. subterranea* is primarily a parasite of potatoes, it can also produce galls on the roots of tomatoes and some other Solanaceae. The organism overwinters in infected potato tubers and can persist for long periods of time in infested soil under cool, moist conditions (Walker, 1969). Spore balls, 19–85 µm in diameter, form a brown dust in the pustules of infected tubers. Each spore bail contains numerous individual, uninucleate resting spores. A resting spore can germinate by releasing a biflagellate zoospore. These motile spores can penetrate roothairs. A thallus is formed which enlarges, forms many nuclei, and eventually segments to produce zoosporangia. Fifty or more zoospores may be produced by each sporangium. Each zoospore typically has two flagella of unequal length. Zoospores with four or six flagella may develop occasionally as a result of fusion (Walker, 1969).

E. Control

Disease-free seed potatoes should always be used for planting. Planting in infested soil should be avoided. Long rotations are helpful. Seed treatment is ineffective. Resistant cultivars are used in some countries, including Germany, Russia, and Chile. There appears to be no good source of resistance among most United States cultivars and seedlings (Manzer *et al.,* 1964a; O'Brien and Rich, 1976; Walker, 1969).

XII. RHIZOCTONIA OR BLACK SCURF

Rhizoctonia is one of the oldest diseases of plants. It causes damping off of seedlings, root and stem cankers of growing plants, and black scurf on potato tubers. It is probably worldwide in distribution.

It is called Rhizoctonia or Rhizoctonia canker after the generic name of the imperfect stage of the causal fungus, rhizoctoniosis, stem canker, sprout canker, black scurf, or "the dirt that won't wash off" in English (O'Brien and Rich, 1976). The French call it *rhizoctonie* (Conners, 1967).

A. Importance

Rhizoctonia is one of the most serious diseases of potato and other crops. It reduces stands, yield, quality, and price. Banville (1978) reported that *R. solani* on seed pieces reduced yields of Norland and Green Mountain by 16–30% and 21–34%, respectively. It also affects size, shape, and appearance of potato tubers (Weinhold and Bowman, 1977). Lack of effective control measures adds to its importance. It is widespread and probably occurs wherever potatoes are grown.

B. Causal Agent

There is some disagreement over the correct scientific terminology for the fungus. The imperfect (mycelial and sclerotial) stage of the fungus is usually cited as *Rhizoctonia solani* Kühn. The latest name for the perfect or basidial stage of the fungus appears to be *Thanatephorus cucumeris* (Frank) Donk (Walker, 1969). Older names or synonyms include *Pellicularia filamentosa* (Pat.) Rogers and *Corticium solani* (Prill. & Del.) Bourd. & Galz. (Conners, 1967; O'Brien and Rich, 1976; Walker, 1969).

C. Symptomatology

Rhizoctonia frequently attacks the sprouts or young shoots below ground level, causing brown cankers on the white stems. These cankers may girdle the stems, causing production of secondary or tertiary stems, resulting in delayed emergence and uneven stands (Fig. 3.15). Affected plants are weak

Fig. 3.15.   Potato stems girdled by *Rhizoctonia solani* below ground level. (Photo courtesy of Maine Life Sciences and Agricultural Experiment Station.)

if severely cankered or girdled. Older plants may exhibit rolling of the leaves, resembling leaf roll, psyllid yellows, or purple top wilt. Aerial tubers (Fig. 3.16) may develop in the axis of the leaves due to interference with starch translocation.

Tubers are often rough, misshapen, and few in number, or numerous very small tubers may result. They often set near the surface, are exposed to light, and turn green. Stolons are frequently girdled resulting in a reduction in yield. Roots may also be attacked.

Fig. 3.16. Potato plant with aerial tubers resulting from girdling of lower stem by *Rhizoctonia solani*. (Photo courtesy of Maine Life Sciences and Agricultural Experiment Station.)

The most conspicuous signs are the numerous small (1 mm or less) to fairly large (more than 1 cm) black sclerotia on the surface of infected potato tubers (Fig. 3.17). They are more conspicuous when tubers are wet, but they are not removed by washing—hence, the name "the dirt that won't wash off." The sterile, septate mycelium is characterized by right-angle branching and a "pinched" appearance where it branches. It is hyaline when young but turns brown with age.

The sexual or basidial stage appears as a white to dirty gray, mat or mycelial felt at the base of the plant near ground level (Fig. 3.18). Basidiospores, 7 to 12 by 1.5 to 3.5 μm, are produced on the basidia which are borne on this mycelium. This stage occurs only under highly humid conditions (Rich, 1977; Walker, 1969).

D. Etiology

Rhizoctonia solani has a very wide host range. It can live on living plants, seeds, plant parts, or in the soil as active mycelium or in the sclerotial stage. It overwinters on potato tubers as black sclerotia of variable size. Infected seed potato tubers are shipped from one area to another and are planted in infested or noninfested soil. Thus it is spread readily from one place to another.

Fig. 3.17. *Rhizoctonia solani* sclerotia on surface of potato tuber.

Fig. 3.18. *Thanatephorus cucumeris,* perfect stage of *Rhizoctonia solani,* on potato stem.

The temperature range for growth of the fungus in culture is 8° to 35°C with an optimum of 25° to 30°C. Sclerotia germinate between 8° and 30°. The optimum temperature for sclerotial germination is 23°C and for basidiospore germination is 21° to 25°C. The optimum temperature for development of stem lesions on potato is about 18°C (Walker, 1969).

Apparently there are numerous pathogenic races of *R. solani.* Isolates from potato may or may not be pathogenic on other crops, such as sugar beets and cabbage, while isolates from these crops may or may not parasitize potato. Isolates from sclerotia on potato also vary widely in their pathogenicity to potato (Walker, 1969). The most common race on potatoes in the United States is *R. solani* anastomosis group (AG) 3 (Davis, 1978; Frank, 1978; Weinhold *et al.*, 1978). Soil-borne inoculum appears to be more important than tuber-borne inoculum in perpetuating this disease (Frank, 1975; 1978).

Frank and Francis (1976) found that *R. solani* produces a phytotoxin capable of causing disease symptoms (root necrosis, stolon pruning, leaf curling, stunting, and chlorosis of leaf margins). Highly susceptible plants were killed within a week. They developed a technique to evaluate resistance to the root necrosis phase of the disease.

E. Control

There are no really good control measures for Rhizoctonia which are both practical and effective. Soil disinfestation is an effective control measure for damping-off of seedlings, but usually it is not practical for potatoes. Crop rotation may have some beneficial effect, but the fungus has such a wide host range and is so easily reintroduced as sclerotia on seed potatoes that it is not very effective. Chemical treatment of seed potatoes with mercury compounds to kill the tuber-borne sclerotia was recommended and practiced for many years. However, this did not control the soil-borne inoculum and proved to be rather ineffective. The use of mercury fungicides is now prohibited in most countries. We often turn to resistant cultivars, but commercially resistant cultivars are unavailable.

Green sprouting or planting seed pieces with short, stubby, actively growing sprouts should prove beneficial. Shallow covering of seed pieces is one of the most important and most practical control measures. This allows the fungus less opportunity to attack the susceptible sprouts. It is a common practice among growers to level off the tops of the rows by dragging a plank, chain, or spike tooth harrow over them, thus permitting fairly deep planting and shallow covering. Soil is later worked in around the base of the plants after emergence to control weeds and prevent greening of the new tubers. Crop rotation with cereals and grasses is beneficial in reducing Rhizoctonia and other soil-borne diseases. Frank and Murphy (1977) suggested a 2-year rotation of oats and potatoes. In their trials, corn, oats and soy beans were poor substrates for *R. solani,* while buckwheat and sugar beets were good substrates. Easton (1978) found that alfalfa and red clover favor the Rhizoctonia disease. Soil treatment with pentachloronitrobenzene is suggested for trial (Easton, 1978; O'Brien and Rich, 1976; Rich 1977). Davis (1973) and Bolkan (1976) obtained beneficial results from seed treatment with benomyl, while Edgington and Busch (1967) observed that oxathiin reduced disease symptoms. However, Prank (1975) found that soil-borne inoculum is more important than tuber-borne inoculum. Apparently some progress is being made toward breeding resistance to this disease. Frank *et al.* (1976) developed a method for rating clone reaction to *R. solani.* Gronquist and Anderson (1977) also described a greenhouse test for evaluating the response of potato cultivars to *R. solani.* They rated some common cultivars as either susceptible or moderately susceptible, but none of those tested was classified as resistant. Banville (1978) demonstrated tolerance to the disease (canker phase) in some cultivars, but there was no evidence of resistance to the black scurf stage.

XIII. SILVER SCURF

Silver scurf is a common disease of potato tubers, but usually it is not too serious. Its geographic range includes Africa, Asia, Australia, Europe, North America, America, and South America.

The most common English name for this disease is silver scurf. The French call it *tache argentee* or *gale argentee de la pomme de terre.* Spanish names include *costra plateada, mancha plateada,* and *caspa plateada. Silberfecken* is the German name for it (Conners, 1967; Miller and Pollard, 1976).

A. Importance

This disease is of relatively minor importance most of the time. It can cause tubers to shrivel during storage, and thus reduce their grade and marketability. It can be a problem if potatoes are grown in muck soils. The practice of washing potatoes has increased the importance of this disease because it is more noticeable on washed potatoes.

B. Causal Agent

Much of the recent literature designates the causal agent as *Helminthosporium solani* Dur. & Mont. (Conners, 1967; Western, 1971). Other writers still call it *Spondylocladium atrovirens* Harz (Miller and Pollard, 1976; Weber, 1973). *Helminthosporium atrovirens* (Harz) Mason & Hughes is another synonym found in the literature. This fungus is a member of the Fungi Imperfecti.

C. Symptomatology

Symptoms of this disease are confined to the tubers. The most obvious symptom is the development of a smooth, gray, leathery skin, especially near the stem end of affected tubers (Fig. 3.19). Symptoms can develop prior to harvest or in storage. They are more conspicuous when tubers are wet, at which time they exhibit a silvery sheen, hence the name "silver scurf." Symptoms are more pronounced on white-skinned cultivars. However, it may destroy the color of red-skinned tubers. Sloughing of the outer skin may occur. Severely affected tubers shrivel and shrink due to loss of moisture (Huguelet, 1976; Miller and Pollard, 1976; O'Brien and Rich, 1976).

Hyphae are hyaline to brown, septate, and branched. They are 1–5 μm wide. Conidiophores are brown, erect, septate, 150–350 × 7–8 μm. Conidia

Fig. 3.19.   Silver scurf, caused by *Helminthosporium solani,* on potato tubers.

are brown, arranged in whorls, 2–8 septate, 24–85 × 7–11 μm at broadest point with tapered tips (Walker, 1952; Weber, 1973; Western, 1971).

Silver scurf can be diagnosed by placing suspected tubers in a moist chamber for several days. If tubers are infected, characteristic dark conidiophores and conidia develop which can be seen easily with the aid of a hand lens (Huguelet, 1976).

D. Etiology

Silver scurf is a disease of potato tubers. The fungus overwinters in tubers in storage or in the field. Huguelet (1976) indicates that it can also overwinter in the soil. It is favored by high soil moisture and by high humidity in storage. The temperature range for fungus growth is 2°–31° with an optimum between 21° and 27°C. Cultivars vary in susceptibility, but none are known to be resistant.

E. Control

Potatoes should be harvested as soon as they are mature. If they remain in moist soil, severity of the disease will increase. Affected tubers should be marketed at once or stored at low humidity. Disease-free seed should be planted. Crop rotation will reduce the inoculum and thus reduce the prevalence and severity of disease. Planting early maturing and russet-skinned cultivars is suggested. Highly susceptible cultivars should be avoided. Seed treatment appears to be ineffective (Hodgson et al., 1974; Huguelet, 1976; O'Brien and Rich, 1976). Soil treatment with pentachloronitrobenzene may be beneficial (Wright, 1968).

XIV. VERTICILLIUM WILT

Verticillium wilt of potato is a serious disease, especially in cool climates. In the past it has often been confused with Fusarium wilt and other wilt diseases. Other English names for this disease include early maturity, early dying, potato wilt, and Verticillium hadromycosis. French-speaking people call it *fletrissure verticillienne, maladie jaune,* or *verticilliase.* Spanish names for it include *marchitez de la papa, verticilosis, Verticillium hadromycosis,* and *tizón.* Germans usually call it *Welkekrankheit* (Miller and Pollard, 1976).

A. Importance

According to Orton (1914), Verticillium wilt was a common disease in Germany and the northern part of the United States as early as 1914. However, not a great deal of attention was paid to it until about 1950. Folsom *et al.* (1951) noted that losses of up to 50% have been reported in Maine. In 1951, Folsom (1953) observed fields with 60–80% of the plants affected in Aroostook and Piscataquis counties in Maine.

McKay (1926) reported that, contrary to general belief, *V. albo-atrum* is more serious than *Fusarium oxysporum* in western Oregon. Yield reductions of 20 to 25% are common in western United States (McLean, 1952). Inoculation with *V. albo-atrum* reduced yields 8000 lbs/acre in Ontario, Canada (Busch, 1965). Recent reports indicate that this disease is also a problem in Australia and Israel (Campbell and Griffiths, 1973; Susnoschi *et al.*, 1975).

B. Causal Agents

Verticillium wilt was first attributed to *Verticillium albo-atrum* by Reinke and Berthold in Germany in 1879 (Westcott, 1971). It is a member of the Fungi Imperfecti. Conners (1967) states that for many years Verticillium wilt was not distinguished from Fusarium wilt, but that when they were recognized as two distinct diseases the causal agent of Verticillium wilt was identified as *V. albo-atrum*. This fungus was also identified as the causal agent of the early maturity disease in Oregon, early dying in Idaho and Washington, and Verticillium wilt in Maine (Folsom *et al.*, 1955; Young and Tolmsoff, 1957).

Some investigators recognize two distinct morphological types of *V. albo-atrum*, while others separate them into two distinct species. Robinson *et al.* (1957) found that isolates from potatoes grown in Wisconsin and eastern Canada had only a dark, resting type of mycelium (DM). Cultures from Oregon and Idaho, on the other hand, produced pseudosclerotia (PS). They believe that the former is indigenous to Maine and the latter is indigenous to Idaho. They classified the DM type as *V. albo-atrum* Reinke and Berth. and the PS type as *V. dahliae* Kleb and this classification has been adopted by most research workers in recent years. *Verticillium albo-atrum* is the dominant species in cool climates, such as Maine (Folsom *et al.*, 1955), Wisconsin, and eastern Canada (Beckman, 1973; Robinson *et al.*, 1957). *Verticillium dahliae* occurs more frequently in slightly warmer climates, such as Idaho, Oregon, Rhode Island, and Connecticut (Edgington, 1962). Both species are present in Ontario, Canada, and Victoria, Australia, but *V. albo-atrum* is more pathogenic (Busch, 1966b; Campbell and Griffiths, 1973).

MacGarvie and Hide (1966) isolated *Verticillium* spp. from 79% of 225 potato seed lots in Great Britain. They identified 72.5% of the isolates as *V. tricorpus*, 10.2% as *V. nubilum,* and 8.5% as *V. nigrescens.* Neither *V. albo-atrum* nor *V. dahliae* was recovered from any of these seed stocks.

"Pink eye" or "brown eye" of potato tubers appears to be associated somehow with Verticillium wilt. However, the exact relationship is not fully understood. Bacteria in the genus *Pseudomonas* have been incriminated (see Pink Eye, Chapter 2, Section VI). Apparently they infect the tubers after Verticillium wilt has predisposed them to attack (Frank *et al.,* 1973).

C. Symptomatology

Plant symptoms usually start to develop at about blossom time. Early symptoms involve epinasty and wilting or "flagging" of the lower leaves. Gradually the leaves turn dull green, then yellow (chlorotic) and finally brown (necrotic). Symptoms progress upward until the entire stem is affected. It is not uncommon for one or more stems in a hill to develop severe symptoms, or even die, while adjacent stems remain symptomless (Folsom, 1953).

Under cool temperatures and high moisture conditions, especially late in the growing season, infected plants may not actually wilt, but they turn yellow, wither, and die from the base upward (Fig. 3.20). It is easy to confuse plant symptoms with those of black leg, ring rot, southern bacterial wilt, Rhizoctonia, and Fusarium wilt. However, the xylem or internal woody portion of a stem infected with *Verticillium* develops a tan or light brown color (Fig. 3.21). This symptom can be observed best if the stem is peeled or cut at a sharp angle at about ground level. A light brown discoloration of the vascular ring of infected tubers can be seen if the stem end is removed. The discoloration is lighter than that associated with *Fusarium.* The vascular ring does not break down into a characteristic rot associated with ring rot or southern bacterial wilt. However, one must take care not to confuse the vascular discoloration associated with chemical vine killers or frost with Verticillium wilt. It may be necessary to isolate the fungus before positive identification can be made (Hodgson *et al.,* 1974; Huguelet, 1976; Rich, 1977, 1978; Robinson *et al.,* 1957).

It is not uncommon for a pinkish or tan discoloration to develop around the eyes or as blotches on the surface of light-skinned tubers produced by plants with wilt symptoms. This symptom has been termed "pink eye" or "brown eye." Under relatively dry storage conditions, the shallow lesions turn brown, dry up, and little attention is paid to them. However, if they do not dry up, secondary or tertiary organisms may invade the tubers and cause them to rot. The Kennebec and Sebago cultivars are most susceptible

Fig. 3.20.   Potato field showing Verticillium wilt symptoms.

to this syndrome. It is sometimes mistaken for late blight tuber rot. Kennebec tubers may develop an internal breakdown, if stored under highly humid conditions (Folsom *et al.,* 1955; Frank *et al.,* 1973; Hodgson *et al.,* 1974; Rich, 1977; Robinson *et al.,* 1957).

Verticillium albo-atrum produces septate dark-colored mycelium. *Verticillium dahliae* produces a lighter colored septate mycelium and pseudosclerotia or microsclerotia. The mycelium and pseudosclerotia are often visible on dead or dying potato vines. Conidia of both species are one-celled, hyaline, globose to ellipsoid, formed at tips of whorled branches, and separate readily from tips. The typically verticilliate or whorled branches are responsible for the genus name, *Verticillium* (Robinson *et al.,* 1957; Westcott, 1971). Western (1971) gives a more complete description of the two fungi.

D. Etiology

The pathogens responsible for Verticillium wilt have a wide host range, including tomatoes, eggplants, strawberries, raspberries, and many species of trees. Numerous common weeds are suscepts also, including *Chenopo-*

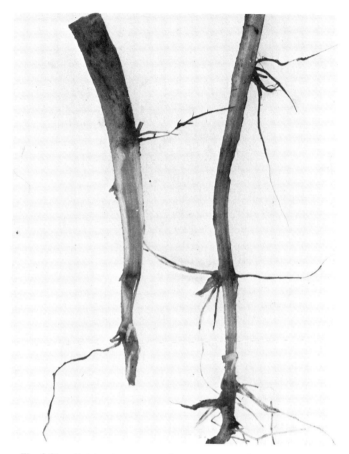

Fig. 3.21. Potato stems cut to show symptoms of Verticillium wilt.

dium album L., *Capsella bursa-pastoris* (L.) Medik, *Taraxacum* spp., and *Equisetum arvense* L. (Engelhard, 1957; Wooliams; 1966). The amount of soil inoculum is influence by crop rotation. it is greatly reduced when non-susceptible crops, such as cereals, grasses, and certain legumes, are included in the rotation. Conversely, inoculum builds up rapidly in the soil when potatoes are planted every year or rotated with other susceptible crops. Huisman and Ashworth (1976) found that Verticillium persisted for 10 years or more in California soils and still attacked cotton. Inoculum declined somewhat when nonsusceptible crops were grown, but built up rapidly when replanted to cotton. Busch (1973) found that cropping practices had little effect on *V. dahliae* in Canadian soils but that *V. albo-atrum* maintained a high level of inoculum only when potatoes were planted annually.

Some common cultural practices favor the incidence and severity of Verticillium wilt. These include continuous cropping of potatoes and planting highly susceptible cultivars such as Kennebec. Abandonment of chemical treatment of seed potatoes prior to planting and the formerly common practice of burning potato vines also favors the buildup of inoculum.

Inoculum in field soil or in soil adhering to seed tubers is more important than mycelium within the tubers (Robinson and Ayers, 1961; Thanassoulopoulos and Hooker, 1968). Frequently inoculum is distributed from one state or province to another by infested soil adhering to seed potatoes (Beckman *et al.,* 1969; Easton *et al.,* 1972). Propagules can also be disseminated from one field to another by dirty field equipment, irrigation water, or blowing dust.

The presence of nematodes, especially *Pratylenchus penetrans,* favors the development of Verticillium wilt (Morsink, 1966; Morsink and Rich, 1968). Hide and Corbett (1973) also showed a synergistic action between *Globodera rostochiensis* and *Verticillium dahliae.*

E. Control

Insofar as possible, infected and/or infested seed potatoes should be avoided. As this disease is spread from place to place via infested soil adhering to the surface of seed tubers, they should be decontaminated with an effective chemical. Organic mercuries are effective and were recommended at one time (Robinson and Ayers, 1953), but their use has been prohibited in recent years in the United States and some other countries. Captan and metiram are among the recommended and approved fungicides. Liquid treatments are superior to dusts because they remove more of the infested soil (Cetas, 1970; Cole *et al.,* 1972; Easton *et al.,* 1972). One kilogram of active ingredient suspended or dissolved in 500 liters water should be effective. The dipping solution should be discarded and renewed after it becomes dirty.

Crop rotation is an important cultural practice. Potatoes should be grown in rotation with cereals, grasses, legumes or other nonsusceptible crops. Growing these crops has proven effective. Highly susceptible solanaceous crops such as eggplant and tomato should not be included in the rotation. Control of weeds is also important because numerous common weeds are suscepts (Hodgson *et al.,* 1974; Wooliams, 1966).

Metham (sodium methyldithiocarbamate) benomyl, and aldicarb, a systemic insecticide, have shown promise as effective soil treatments (Biehn, 1970; Cetas, 1970; Hoyman and Dingman, 1965, 1967; Miller *et al.,* 1967; Young, 1956; Young and Tolmsoff, 1957). Busch (1966a) tried Di-Syston, another systemic insecticide. These chemicals delayed symptoms and increased yields under experimental conditions. Further work is needed to determine their exact role in commercial potato production.

Soil fumigation with nematicides has proven beneficial, especially when nematodes were prevalent (Easton, *et al.,* 1969). Ethylene dibromide, D-D (dichloropene–dichloropropane), or related compounds can be used for this purpose. However, under severe wilt conditions a combination fungicide–nematicide, such as metham, MIT (Vorlex), or methyl bromide, should prove more effective. A single fall application of Vorlex in Connecticut resulted in increased potato yields for the 3 following years (Miller and Hawkins, 1969). Easton *et al.* (1974) found that 25 gal of Vorlex per acre increased yields 6–8 tons per acre in the state of Washington. Powelson and Carter (1973) also observed that soil fumigation not only delayed symptoms but also increased the value of the potato crop for 2 years after fumigation.

Potato growers should avoid planting susceptible cultivars in soil with a history of Verticillium wilt. These include Arran Consul, Avon, Belleisle, Cherokee, Chieftain, Chippewa, Earlaine, Epicure, Fundy, Irish Cobbler, Kennebec, Keswick, Norchief, Norchip, Norland, Onaway, Red La Soda, Red Pontiac, Russet Burbank (Netted Gem), Sebago, Sioux, Superior, Triumph, Warba, Waseca, Wauseon, White Rose, and York (Anonymous 1964; Ayers, 1961b; Busch, 1965; Folsom *et al.,* 1951; Munro, 1975).

There are many resistant cultivars from which to choose. Some of the older ones are Houma, Hunter, Menominee, Mohawk, Monona, Ona, Ontario, Pontiac, Red Beauty, Russet Rural, Sequoia, and Yampa (Busch, 1966; Folsom *et al.,* 1951; Hodgson *et al.,* 1974; McLean and Akeley, 1957; Munro, 1975; Rieman and Schultz, 1955; Robinson and Ayers, 1953; Robinson *et al.,* 1957). Among the newer wilt-resistant cultivars are Abnaki (Akeley *et al.,* 1971), Cariboo (Maurer *et al.,* 1968), Cascade (Hoyman, 1970), Nooksack (Hoyman and Holland, 1974), Raritan (Campbell and Young, 1970), Seminole (Stevenson *et al.,* 1970), and Targhee (Pavek *et al.,* 1973a). Some cultivars are tolerant but not resistant to Verticillium wilt, and produce satisfactory crops in spite of infection by the wilt organisms. Reliance (Hoyman *et al.,* 1963) and Shurchip (O'Keefe, 1970a) are in this category.

A recent report (Murphy *et al.,* 1982) included a number of additional cultivars and seedlings which are reported to be resistant to Verticillium wilt. The list includes Batoche, Belrus, Campbell 11, Campbell 13, Rideau, Russette, Shepody, Tobique, AF303-5, BR7088-18, BR7093-23, C7358-26A, CA02-7, and CF7353-1.

It is probable that pathogenic races of *V. albo-atrum* and *V. dahliae* occur in nature as is the case with many other pathogenic fungi. The fact that certain potato cultivars react differently to such factors as abundance of inoculum, geographic location, and different isolates of the fungi support this hypothesis (Susnoschi *et al.*, 1975; Waggoner, 1956). This explains why a cultivar is sometimes listed as resistant while at other times under different conditions it may be classified as susceptible. It also adds to the complexity of breeding for disease resistance.

XV. WART

Potato wart is a minor disease in the United States and Canada with the exception of Newfoundland. However, it is serious in many other parts of the world. Other English names for this disease are black wart, black scab, and canker. French-speaking people refer to it as *tumeur verruqueuse, gale verruqueuse, gale noire,* or *maladie verruqueuse.* The Germans call it *Krebs* or *Kartoffelkrebs.* Spanish names include *verreuga, enfermedad de verruga,* and *roña negra* (Miller and Pollard, 1976).

A. Importance

Potato wart occurs in Africa, Asia, Europe, North America, and South America. It is restricted to small areas in Pennsylvania, Maryland, and West Virginia in the United States, but is a destructive disease in Newfoundland, Canada. It is widely distributed and potentially destructive in Europe, and also occurs in South Africa, India, and South America (Lapwood and Hide, 1971; O'Brien and Rich, 1976).

B. Causal Agent

Wart is caused by *Synchytrium endobioticum* (Schilb.) Perc., a member of the Chytridiales. At least 10 pathogenic races of the fungus have been identified (Lapwood and Hide, 1971).

C. Symptomatology

The principal symptom of this disease is the presence of rough, warty outgrowths or protuberances on tubers, stolons, and stems of potato (Fig. 3.22), also occasionally occurring on leaves and flowers. Roots are not affected. The warts are soft, spongy, and more or less spherical, their size varying from that of a pea to large masses covering entire tubers. Their color resembles that of the host tissue. Underground warts are white to light pink when young but become darker with age. Above-ground symptoms,

Fig. 3.22. Potato plant exhibiting symptoms of wart, caused by *Synchytrium endobioti-cum*. (Photo courtesy M. C. Hampson, Research Station, St. John's, Newfoundland, Canada.)

when they occur, consist of yellowish-green, cauliflower-like outgrowths; often no above-ground symptoms are visible. Warts on older tubers may turn black, hence, the names "black wart" and "black scab" (Lapwood and Hide, 1971; O'Brien and Rich, 1976).

The fungus lacks mycelium. It has two spore stages: one is a thin-walled "summer" sporangium, and the other is a thick-walled "winter" sporangium. Both types produce zoospores in the extruded vesicle or sorus. Each zoospore has one flagellum (Lapwood and Hide, 1971).

D. Etiology

Uniflagellate zoospores, which are produced from resting sporangia, swim about in soil water. They encyst and penetrate the epidermal bud cells of tuber eyes by means of an infection peg. A uninucleate thallus develops which enlarges to form a prosorus. A vesicle then develops from this enlarged host cell. A sorus is formed by the passage of the contents of the enlarged cell or prosorus into the vesicle. The fungus stimulates hypertrophy and hyperplasia of neighboring host cells without actually invading them, thus forming a raised rosette of variable size, depending on the degree of stimulation of meristematic activity. The sorus divides repeatedly into several sporangia in which the zoospores develop. The infected cell wall bursts, discharging some of the sporangia which, in turn, eject the zoospores. New infections result from zoospores produced by these summer sporangia or from resting sporangia in the soil. This process continues throughout the growing season, resulting in more hypertrophy, hyperplasia, and rosette cells (Lapwood and Hide, 1971).

Zoospores from summer sporangia may also function as gametes, combine in pairs, and produce zygotes. These diploid zygotes infect the host in the same manner as zoospores. The epidermal cell infected by the zygote and surrounding cells are stimulated to divide rapidly. Thus the fungus becomes buried deeply in abnormal host tissue and large, rough, irregular, warty tumors are formed. The zygote enlarges, develops a thick wall, and eventually produces a uninucleate resting sporangium. These sporgania are released into the soil when a warty tuber decays (Lapwood and Hide, 1971). Potato virus X is sometimes transmitted by *S. endobioticum* spores (Teakle, 1969).

Wart is favored by wet soils. Optimum conditions for germination of resting spores are wet soil conditions and a temperature of 10°–27°C.

The inoculum is distributed from one locality to another via infected seed tubers and is disseminated locally by infested soil, infested manure, and contaminated machinery. Resting sporangia can survive for many years between potato crops. One report indicated a survival period of 30 years, but this is unusual. Experimental hosts include such solanaceous crops as tomato *(Lycopersicon esculentum* Mill.), *Solanum dulcamara* L., and *S. nigrum* L., but natural infection of these plants is not a problem (Lapwood and Hide, 1971).

E. Control

Certified, disease-free seed potatoes should always be used for planting. Quarantines should be strictly enforced. Sanitation to prevent spread of inoculum should be routinely practiced. Long crop rotations will help to minimize losses, and removal of diseased plants should be helpful in reducing the buildup of inoculum. The use of soil fungicides to eradicate the wart fungus is helpful but costly (Hodgson *et al.*, 1974; O'Brien and Rich, 1976).

The principal method of control where wart is a problem is the use of resistant cultivars. Cultivars which are resistant in Europe are not resistant in Newfoundland. Kennebec, Sebago, Pink Pearl, and Urgenta are recommended for use in Canada. In the United States, Irish Cobbler, Katahdin, Mohawk, Sequoia, Mesaba, and Norkota are immune, and Houma is resistant (Hodgson *et al.*, 1974; O'Brien and Rich, 1976).

In the Ukraine the following cultivars are reported to be resistant: Agro, Mira, Gievont, Temp, and Hilla (Saltykova, 1973). Additional cultivars which are highly resistant in Russia include Ekaterininskiĭ, Belouisskiĭ ranniĭ, Imandra, Loshiskiĭ, and Severnaya roza (Bodarenko, 1973). Breeding for resistance is complicated by the tremendous differences in pathogenicity of the various races of the pathogen.

XVI. MISCELLANEOUS FUNGUS DISEASES

There are numerous potato diseases incited by fungi which occur only occasionally or are usually of minor importance. However, they may become important locally when environmental conditions favor their development. A brief discussion of each will be included here.

A. Armillaria Rot

Armillaria rot, caused by the Basidiomycete, *Armillaria mellea* (Vah. ex Fr.) Kummer, is primarily a disease of forest and fruit trees. It is usually referred to as Armillaria root rot or mushroom root rot on trees. The fungus may be referred to as *Armillariella* rather than *Armillaria* in some of the more recent literature. It has been reported on potato tubers in western United States and western Canada (Blodgett and Rich, 1950; Conners, 1967).

External symptoms consist of light brown, slightly sunken, rotted areas on affected tubers. The internal rotted tissue consists of alternating layers of yellowish and white tissue. Signs consist of a few long black strands or rhizomorphs of the fungus tissue connecting the rotted areas of the tubers.

No control measures are usually required for this disease. However, it would be wise to avoid planting potatoes on newly cleared orchard or wood land.

B. Coiled Sprout

This disease has gained attention in Europe in recent years. The exact cause is not fully known, but *Verticillium nubilum* Pethybr. appears to be involved in some way.

Symptoms of the disease include bending (coiling), swelling, splitting, and fasciation of the sprouts. Increased lateral bud growth and lenticular intumescences are also often present. Symptoms start to develop about 4 days after planting. Long sprouts, soil compaction, and deep planting favor the development of coiled sprout.

Verticillium nubilum produces russet brown lesions on the inside of the coil. It can be distributed as superficial infections of seed tubers or in infested soil adhering to the seed tubers.

Control measures include seed disinfestation and use of seed potatoes without long sprouts. Recommended cultural practices are shallow planting and prevention of soil compaction over the seed pieces or sets (Catchpole and Hillman, 1975a,b,c; Lapwood *et al.,* 1977; Western, 1971).

C. Gray Mold

Gray mold is caused by *Botrytis cinerea* Pers. ex Fr., a weakly pathogenic fungus in the Fungi Imperfecti. It attacks both plants and tubers and it is often found on older, senescent leaves, near the ground. Affected leaves turn brown and may be covered with a dark brown or dark gray mold. Sometimes it is confused with late blight, but the late blight fungus produces a white or light gray mold.

The mycelium is septate, hyaline at first but darkens with age. The conidia of *B. cinerea* are hyaline, one-celled, oval, 11–15 × 8–11 μm. Sometimes sclerotia 1–6 mm in diameter are formed (Western, 1971).

Gray mold on tubers is primarily a storage and transit disease. The fungus has a very wide host range and spores are almost omnipresent. They enter tubers through wounds where they sporulate and produce a brown, odorless decay. Infected tubers become rubbery and shrunken (Weber, 1973).

The most important control measure is to avoid wounding the tubers during the harvesting operation. Potatoes should be stored at 18°–20°C for 2 weeks to hasten healing of wounds. The storage temperature should then be lowered to 5°–10°C. Some fungicides used for the control of late blight will minimize infection of the leaves. Captafol is more effective than maneb for control of gray mold (O'Brien and Rich, 1976).

D. Jelly End Rot

Jelly end or jelly end rot is a common defect of long tubers, especially those which are pointed at the basal or stem end. At first the stem end of an affected tuber becomes translucent or "glassy," giving rise to the name "glassy end." This is usually followed by a soft, jelly-like, light brown rot, extending 2.5–3.0 cm from the stem end. Affected tissues usually become dry and shriveled, and a clear line of demarcation develops between affected and healthy tissues.

The cause of jelly end rot is not clearly understood. *Fusarium solani* f. *radicicola* Wr. Snyder & Hansen is frequently isolated from diseased tubers. However, many diseased tubers are apparently sterile. Dry soil conditions toward the end of the growing season appear to favor jelly end rot. Possibly the vines withdraw moisture from the stem end of the tubers. This may increase their susceptibility to weakly pathogenic organisms.

No effective control measures are known. Maintenance of an adequate moisture supply throughout the growing season is important. Cultivars which produce long tubers with pointed ends such as Russet Burbank should be avoided. Round or oval-shaped tubers are rarely if ever affected (Blodgett and Rich, 1950; O'Brien and Rich, 1976).

E. Leaf Blotch

Leaf blotch is a minor disease of potato foliage. Faint, yellowish green, irregular blotches, up to 1.2 cm in diameter, develop on the lower leaves with symptoms gradually spreading to the middle and upper leaves. The spots or blotches become yellow or brown with age, and the lower surfaces of the blotchy areas exhibit a violet gray fungus growth. It has been reported from eastern, midwestern, and southern states in the United States.

The disease is caused by *Cercospora concors* (Casp.) Sacc. *Cercospora solani* Thuem. has also been reported on potato in Alabama and Texas (Anonymous, 1960). The mycelium of *C. concors* is hyaline or slightly colored, septate, and branched. Conidiophores 20–75 μm long and 7 μm wide protrude through the stomata. Conidia are hyaline to light colored, septate, and mostly straight, ranging in size from 20–100 × 4–6 μm.

As the disease is of only minor importance, no specific control measures are recommended. Probably foliar fungicides would reduce the severity of the disease (Chupp, 1953; Weber, 1973).

F. Penicillium Rot

Penicillium causes a mold or rot of many fruits, grains, and vegetables, including potato tubers and seed pieces. The *Penicillium* blue mold fungus grows on skinned, wounded, or cut surfaces. It is capable of growing at temperatures as low as 0°C. It is characterized by blue-green spore masses on injured tubers. Rotted tissue beneath the mold is moist, fairly firm, and has a moldy odor. This disease may be a problem when injured potatoes or cut seed pieces are stored near 0°, since this low temperature is not conducive to healing or suberization of the cut or injured surfaces but permits the fungus to become established and continue to grow. The exact species determination of the fungus is uncertain. It is listed as *Penicillium* sp., in part *P. oxalicum* Currie & Thom (Anonymous, 1960; Blodgett and Rich, 1950; Chupp and Sherf, 1960).

Potatoes should be handled carefully to prevent injury. They should be stored at 12°–16°C for a few days to hasten suberization. Prolonged storage at near 0°C and high humidity should be avoided (Blodgett and Rich, 1950; A. E. Rich, unpublished data).

G. Powdery Mildew

This disease is usually not serious on potato, although Polli and Moeller (1944) reported that it caused epiphytotics in Palestine for 3 consecutive years. It reduces yield by 20% or more and lowers grade when it occurs early in the season. Thomas (1946) observed it in the greenhouse, trial plots, and neighboring fields in England. Mycelium was more prevalent on the upper surface of the leaves in the greenhouse and on the lower surface in the fields. Round to oval patches 1–3 cm in diameter developed and sometimes coalesced. Petioles and stems remained unaffected. Subglobose to ovate haustoria, 41.5×27.5 μm, developed in the epidermal cells. Conidia measured 29.6×16.8 μm. No cleistothecia were observed. Menzies (1950a) first reported the natural occurrence of the perfect stage of *Erysiphe cichoracearum* DC. on potatoes and stated that the mildew was inconspicuous with sparse mycelial and conidial production. It was confined to terminal growth of maturing vines and was most common on petioles and stems producing a light superficial russetting of affected tissues. The disease occurred so late in the growing season that little or no damage resulted from it. Rowe (1975) observed powdery mildew in Ohio and tentatively identified it as *E. cichoracearum*. Chupp and Sherf (1960) cite the causal agent of powdery mildew in Palestine, Turkey, and other dry countries bordering the Mediterranean Sea as *Leveillula taurica* (Lev.) Arnand.

Control measures are usually not necessary. Suggestions include the use of resistant cultivars such as Ulster Torch and dusting with sulfur (Dutt *et al.*, 1973; O'Brien and Rich, 1976; Polli and Moeller, 1944).

H. Rusts

Two rusts on potato have been reported from Latin America. *Aecidium cautensis* Arth. was collected by Abbott in Peru occurring at 8000 to 9000 ft elevation. It also attacks tomatoes. *Puccinia pitteriana* P. Henn. is a mycrocyclic rust. Although it causes some defoliation in Mexico and Peru, it is not widespread or frequently observed. It attacks *Solanum demissum* in Mexico. Very little information is available concerning these two rusts (Walker, 1952; Weber, 1973).

I. Skin Spot

This disease has been reported occasionally in the United States and Canada, but is not usually considered to be a problem. However, it has become a serious problem in Great Britain in recent years. All underground parts of the plant are affected. Light brown lesions develop on stems, stolons, and roots. Lesions are usually larger and lighter colored than lesions caused by *Rhizoctonia solani*. The lesions coalesce, crack transversely, and the affected cortical tissue becomes loosened. Purplish black pimples, 0.5–2.0 mm in diameter with raised centers, develop on stored tubers (Fig. 3.23). Tuber eyes are also affected. When such tubers are used for seed, they sprout unevenly if at all, resulting in an uneven stand with many missing hills.

Skin spot is caused by the fungus *Oospora pustulans* Owen & Wakef., a member of the Fungi Imperfecti. The mycelium is hyaline, septate, with short, erect, branched conidiophores arising from it. Conidia are nonseptate, hyaline, 2–2.5 × 6–12 μm. Tubers can become infected during growth, during harvest, or in storage. The fungus enters through lenticels or superficial wounds. Black sclerotia, 49 μm–1 mm, are produced in aging cul-

Fig. 3.23. Skin spot of potato tubers caused by *Oospora pustulans*. (Photo courtesy of Maine Life Sciences and Agricultural Experiment Station.)

tures and in decayed tuber tissue. Infection usually results from diseased seed tubers, but sclerotia can also serve as a source of inoculum (Anonymous, 1960; Boyd and Lennard, 1961; Conners, 1967; Hodgson *et al.,* 1974; Lapwood and Hide, 1971; Walker, 1952).

Suggested control measures include use of disease-free seed. Use of stem cuttings to produce disease-free seed should be beneficial. Other recommended practices are crop rotation, early harvesting, harvesting when dry, and storage in crates instead of in bulk. Seed treatment with mercury was recommended at one time, but this practice is no longer permissible. Although some cultivars such as King Edward are more susceptible than others, none seems to be resistant (Allen, 1957; Bannon, 1975; Boyd, 1957; Boyd and Lennard, 1961; Hide *et al.,* 1969; Hodgson *et al.,* 1974; O'Brien and Rich, 1976; Salt, 1964).

J. Smuts

A smut disease on potato was found in Peru by Barrus (1944) and Barrus and Muller (1943). Barrus named the smut fungus *Thecaphora solani.* Although this name has been continuously used by most plant pathologists, recently, O'Brien and Thirumalachar (1972) erected a new genus and classified it as *Angiosorus solani.* They noted that the disease occurs widely in Peru, Ecuador, and Venezuela but is confined to South America. It is sometimes called gangrene, or *gangrena* in Spanish. Bazan De Segura (1960) reported losses of up to 80% on the cultivar Peruanita. Later he and Del Carpio (1974) reported losses of up to 50% on the Peruvian coast and stated that all commercial cultivars are susceptible. They also noted that high populations of the root knot nematode, *Meloidogyne incognita,* favored disease development.

The disease is expressed by the development of warts or lumps on the surface of the leaves or tubers with internal brown specks (Fig. 3.24), cavities of which are filled with rusty brown spore balls. The spore balls are ovoid, consisting of 1-8 spores, covered with verrucose markings, and measure 12-48 × 12-35 μm. Individual spores are 7.5-20 × 8-18 μm and are dusty when dry (Weber, 1973).

Recommmended control measures include use of healthy seed potatoes, crop rotation, burning of infected plants debris, disinfestation of tools, and breeding of resistant cultivars (Bazan De Segura and Del Carpio, 1974). According to Zachmann and Baumann (1975), the cultivars Mariva and Seperia are resistant.

Weber (1973) also includes another smut fungus on potato, listing *Polysaccopsis hieronymi* (Schroet.) P. Henn. as the causal agent. Very little is known about this disease.

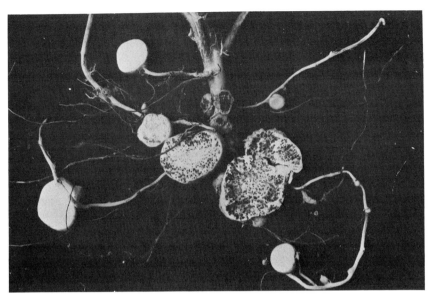

Fig. 3.24. Cross section of potato tubers infected with smut caused by *Thecaphora solani.* (Photo courtesy of Rainer Zachmann, International Potato Center, Lima, Peru.)

K. Southern Blight

This disease is primarily a problem of subtropical and tropical regions, including the southern states of the United States, hence its name southern blight. It is also known as Sclerotium rot.

The causal agent is usually cited as *Sclerotium rolfsii* Sacc. (Chupp and Sherf, 1960; O'Brien and Rich, 1976). However, Walker (1952) states that it is incited by *Pellicularia rolfsii* (Curzi) West, and Weber (1973) classifies it is *Botryobasidium rolfsii* (Sacc.) Venkat., the perfect stage of *S. rolfsii.* This fungus attacks more than 200 species of plants, including most vegetables. It is often girdles the main potato stem at ground level. At first, infected plants wilt during the day and recover at night. As the disease progresses, however, leaves become yellow, turn brown, and die. Sunken, discolored, water-soaked lesions develop on infected stems just below ground level. Later they become covered with a white, threadlike fan-shaped mycelial mat of the fungus. Small spherical resting bodies called sclerotia, measuring 1 mm or less in diameter, develop on the mycelium. They are white at first but gradually turn dark brown. Like *Rhizoctonia,* this fungus does not produce conidia, Basidiospores are rarely found; they are hyaline, one-celled, and 5–10 × 3–6 μm. The fungus can produce a seed-piece decay and tuber rot in addition to causing wilting and death of plants. The rot is

white and practically odorless in early stages of development, gradually be-
coming yellow in later stages. In mild attacks only slightly depressed areas
are produced, but often the decay develops rapidly, and the entire tuber
becomes slimy. This condition may develop during storage and transit
(Chupp and Sher, 1960; O'Brien and Rich, 1976; Walker, 1952; Weber,
1973).

Because this disease causes only minor losses in most potato-growing
areas, little emphasis has been placed on its control. Disease-free seed potatoes, crop rotation with nonsusceptible crops, and weed control are suggested. Potatoes should not be harvested and packed during wet weather
and should be dried thoroughly before packing and shipping.

L. Stalk Rot or White Mold

Stalk rot occurs spasmodically in the United States, Canada, and other
potato-growing areas. It is known by a variety of names including cottony
rot, stalk break, stalk rot, white mold, Sclerotinia white mold, and Sclerotinia disease. The causal agent is usually listed as *Sclerotinia sclerotiorum*
(Lib.) D By. (Conners, 1967; O'Brien and Rich, 1976; Walker, 1952;
Weber, 1973). However, Korf and Dumont (1972) renamed it *Whetzelinia
sclerotiorum;* this name is found in some of the most recent literature. It
is believed that *S. sclerotiorum* does not parasitize potato tubers but, according to Ramsey (1941), *S. intermedia* Ramsey can cause an extensive
white, watery soft rot. *Sclerotinia minor* Jagger is also pathogenic at high
storage temperatures.

Sclerotinia sclerotiorum has a very wide host range, including beans, celery, crucifers, cucurbits, tomatoes, lettuce, and other greens. Symptoms on
potato plants include a light-colored watery soft rot, often confined to the
the main stems. Infected plants eventually wilt and die. Macroscopic signs
consist of a white, cottony mold growing over the surface of infected stalks
(Fig. 3.25). Black sclerotia up to 1 cm long are produced in the hollowed-
out, diseased stems. The fungus survives from one season to the next as
sclerotia in the soil and plant debris. The sclerotia germinate in the spring,
produce apothecia, asci, and ascospores. The ascospores are hyaline, oval,
one-celled, $11–15 \times 5–8 \ \mu m$. True conidia are lacking, but microconidia
are formed in chains (Chupp and Sherf, 1960; O'Brien and Rich, 1976;
Walker, 1952, 1969; Weber, 1973).

Growth of this fungus is favored by relatively cool, moist conditions. It
rarely occurs in the tropics but can be a problem in semitropical areas, such
as Florida, especially during the cooler part of the growing season. It is
favored by poorly drained soils, continuous cropping to susceptible crops,
sprinkler irrigation, and rank growth of plants which provide shade and

Fig. 3.25. Potato stems infected with stalk rot caused by *Sclerotinia sclerotiorum.*

high humidity. The author has observed that mechanical damage to plants, such as injury from sprayer wheels or hail, favors the development of this disease.

Control measures include sanitation, crop rotation, and judicial use of irrigation. Flooding for 5 weeks has been practiced in some instances, but this is impractical for most regions. Soil treatment with pentachloro-nitrobenzene (PCNB, Terraclor) and spraying or dusting with a carbamate fungicide, such as ziram, zineb, or maneb, have been suggested for some crops. However, these practices would be impractical for potatoes (Chupp and Sherf, 1960; O'Brien and Rich, 1976; Walker, 1969).

M. Stem-End Hard Rot

This minor disease of potatoes has been reported from British Columbia and western Washington. The causal agent is *Phomopsis tuberivora* Gussow & W. R. Foster (Blodgett and Rich, 1950; Conners, 1967; Weber, 1973). A hard, dry, corky rot develops at the stolon end of affected tubers and a conical decay progresses toward the center of the tuber. Small tubers may become mummified. Weber (1973) also mentions damage to the leaves but he does not describe it.

The fungus produces black, globose pycnidia. Pycnidiospores are hyaline, one-celled, spindle-shaped, 10–12 × 4–6 μm. The ascospore stage may be *Diaporthe tulasnei* (Weber, 1973).

Potatoes should be stored at 2°–4°C to minimize storage losses. Growers should plant disease-free seed.

N. Stemphylium Leaf Spot

Wright (1947) described a leaf spot on potato in British Columbia caused by *Stemphylium consortiale* (Thuem.) Groves & Skolko. The spots on the leaves resemble early blight lesions but they are lighter brown and lack the characteristic concentric zones caused by *Alternaria solani*. *Stemphylium consortiale* is also pathogenic on tomato.

O. Violet Root Rot

This disease has a wide host range but usually it is of minor importance on potato. It is widely distributed through most of the potato-growing areas of the world. The English, French, Spanish, and German names for it are violet root rot, *rhizoctone violet, pudrición radicular violeta,* and *violetter Wurzeltöter,* respectively.

Above-ground symptoms on potatoes are not distinctive. At first, foliage may become chlorotic, followed by wilting and dying. The fungus attacks the roots and other underground plant parts. When a diseased plant is pulled, the underground parts are covered with a superficial violaceous mat of the fungus. Small, violet-colored lesions are often abundant at the stem end of tubers. The lesions may enlarge and increase in number until the entire tuber is involved. Decayed tissue becomes ashy-gray, dry, and powdery.

The imperfect stage of the causal fungus is *Rhizoctonia crocorum* (Pers.) DC. The perfect stage is a Basidiomycete known as *Helicobasidium purpureum* (Tul.) Pat. Young hyphae are light violet, branched at right angles, with septa not more than 10 μm from the junction. The mycelium develops a dense mass, sometimes producing visible strands. Dark violet sclerotia form on the strands, either on the host root or in the soil. The hymenium of the basidial stage is feltlike and purplish in color. The basidium is hyaline and curved. Two to four unicellular, hyaline single-celled basidiospores, 10–12 × 6–7 μm develop on each basidium. The optimum temperature for fungus growth is 25°C, and for infection is 16°C.

Control measures are not usually required. Where necessary, care should be taken to use disease-free seed potatoes. Long rotations with cereals and grasses along with control of susceptible weed hosts are also suggested (Chupp and Sherf, 1960; Hodgson *et al.*, 1974; Miller and Pollard, 1976; Walker, 1952, 1969; Weber, 1973; Western, 1971).

P. Xylaria Tuber Rot

This disease has been reported on potato tubers in Florida. Light colored, sunken, circular areas of dry rot first appear on the tubers, followed by a soft, cheesy decay which may involve the entire tuber.

The causal agent, *Xylaria apiculata* Cke., like *Armillaria,* is usually a parasite of tree roots. However it is an ascomycete. The septate mycelium is dark and branched. Cream-colored to black stromata, up to 8 cm × 3 mm are formed. Light brown, one-celled, oval to elliptical conidia measure 6–7 × 2–3 μm. Brown, elliptical, one-, or two-celled, ascospores, measuring 16–25 × 4–7 μm, are formed in the asci within the perithecia (Walker, 1952; Weber, 1973).

Potatoes should not be planted on newly cleared land. Tree stumps and roots should be removed from cleared fields.

Q. Other Fungi

A number of additional fungi have been reported to occur rarely on potatoes. Very little information is available concerning their association with potatoes. They include *Aspergillus niger* v. Tiegh., *Ascochyta lycopersici* Brunaud, *Botryosporium longibrachiatum* (Oud.) Maire, *Clonostachys araucariae* Cda. var. *rosea* Preuss., *Cylindrocarpon magnusianum* Wr., *Fusarium* spp., *Gliocladium* sp., *Gloeosporium* sp., *Hypomyces ipomoeae* (Halst.) Wr., *Mycosphaerella solani* (Ell. & Ev.) Wr., *Nectria* spp., *Papulaspora coprophila* (Zukal) Hotson, *Phymatotrichum omnivorum* (Shear) Dug., *Physarum cinereum* (Batsch) Pers., *Ramularia solani* Sherb., *Stysanus stemonitis* Corda, and *Trichothecium roseum* Lk. ex Fr. For further information on these rare, minor, or questionably parasitic diseases of potato, see Anonymous (1960), Conners (1967), or Weber (1973).

Diseases Caused by Viruses,
Viroids, and Mycoplasmas

I. INTRODUCTION

Virus diseases of potato have been recognized for centuries. However, very little was known about the nature of the causal agents until the twentieth century. An early virus disease, or combination of virus diseases, was known as "the curl," "degeneration," "senility," or "running out." Symptoms included leafrolling and/or roughening of the leaves. It may have been a combination of leafroll, rugose mosaic, and other virus diseases. It was thought that the problem was due to continued asexual propagation of the potato, causing it to lose its vigor or "run out." Plants became small and weak, and yields were considerably reduced. It was shown to be tuber-transmitted as early as 1778. It is true that most, if not all, potato viruses are tuber-perpetuated through clonal propagation, thus encouraging the build-up of viruses in a stock unless effective disease control practices are followed. Folsom and Bonde (1936) prepared a list of known potato viruses. Fortunately, potato viruses are not transmitted through the true seed. Therefore, every new potato cultivar produced from seed should be virus-free until it becomes infected. Unfortunately, however, it seems that most new cultivars become infected to some degree during the testing and increase processes before they are ever released to growers.

Potato seed certification programs have been developed in most potato-growing areas in an attempt to assure that cultivars remain true to name and to keep virus infection at a minimum. These programs have been relatively successful.

Viruses are submicroscopic in size and can be seen only with the aid of an electron microscope. They are straight or flexuous rods or polyhedral (sometimes called spherical). Most plant viruses contain ribonucleic acid (RNA) surrounded by a protein coat or capsid (Bawden, 1964). Roberts and Boothroyd (1972) define a virus as an ultramicroscopic (one dimension less than 200 nm), obligately parasitic, and infectious pathogen of disease consisting of a core of nucleic acid in a protein coat.

According to Rhodes and Van Rooyen (1968) true viruses possess the following characteristics.

1. They consist of deoxyribonucleic acid (DNA) or ribonucleic acid (RNA), but not both.
2. A protein shell or tube surrounds and protects the centrally located nucleic acid core.
3. They multiply only in living cells.
4. They do not divide by binary fission.
5. The nucleic acid of the infecting virus takes over the control of the infected cell.
6. They make use of the ribosomes of the invaded host cell.
7. The complete viral particle is known as the virion. It consists of a nucleic acid core surrounded by an antigenically specific capsid or protein coat.
8. The virion may or may not have an envelope outside the capsid.

The nomenclature of viruses is complicated and confusing. The Johnson system gave each virus from a given host or suscept a number, e.g., potato virus 1, potato virus 2. K. M. Smith (1957) used a similar system but used the generic name of the host, e.g., *Solanum* virus 1, *Solanum* virus 2. F. O. Holmes (1948) devised a Latin binomial system with generic and species names similar to that used for classifying living organisms. This system has certain advantages and disadvantages, but it was not widely adopted. Numerous committees have been established to develop a uniform, acceptable system of virus nomenclature, but none has been very successful. A common procedure is to designate the virus by the common name of the disease on the first host on which it was described, e.g., potato leafroll virus. Abbreviations, such as PLRV for potato leafroll virus, are usually used for convenience (Walker, 1969).

A cryptogram system, describing the known properties of a virus, has been developed. The symbols, respectively, refer to: type of nucleic acid in virus particle/strandedness of nucleic acid: molecular weight of nucleic acid in millions/percentage of nucleic acid in virus particle: shape of particle outline/shape of nucleocapsid: kind of host/kind of vector. For example, the cryptogram for potato virus X is R/1:*/6:E/E:S(Fu). This means that potato virus X contains RNA which is single-stranded, has unknown molecular weight and constitutes 6% of the virus particle; both particle outline and nucleocapsid have parallel sides with ends not rounded; the virus infects seed plants, and it may have a fungus vector. The parentheses indicate that fungus transmission has not been confirmed (Harrison, 1971).

A further explanation of the symbols is given in the tabulation below.

Type of nucleic acid	R, RNA; D, DNA
Strandedness	1 = single, 2 = double
Molecular weight of nucleic acid	Molecular weight in millions
Percentage of nucleic acid	Percentage of nucleic acid in infective particles
Outline of particle	S, Essentially spherical
	E, Elongated with parallel sides, ends not rounded
	U, Elongated with parallel sides, end(s) rounded
	T, Having a tail
	X, Complex or none of above
Shape of nucleocapsid	S, Essentially spherical
	E, Elongated with parallel sides, ends not rounded
	U, Elongated with parallel sides, end(s) rounded
	X, Complex or none of above
Hosts	V, Vertebrate
	I, Invertebrate
	S, Seed plant
	F, Fungus
	B, Bacterium
	A, Actinomycete
Vector	O, Spreads without a vector; no vector known
	A, Aphid (Insecta, Hemiptera, Aphididae)
	B, Beetle (Insecta, Coleoptera)
	C, Mealy bug (Insecta, Hemiptera, Coccidae)
	D, Fly (Insecta, Diptera)
	E, Mite (Arachnida, Eriophyidae)
	F, Fungus (Fungi)
	G, Tingid bug (Insecta, Hemiptera, Gymnocerata)
	I, Tick (Arachnida, Ixodiodea)
	L, Leaf-, plant-, or treehopper (Insecta, Hemiptera, Auchenorrhyncha)
	N, Nematode (Nematoda)
	S, Flea (Insecta, Siphonaptera)
	T, Thrips (Insecta, Thysanoptera)
	W, White-fly (Insecta, Hemiptera, Aleyroidae)

The cryptograms give valuable information in a concise form, but often much of the information is unknown, which is indicated by an asterisk. For example, the cryptogram for the potato leafroll virus is */*:*/*:S/S:S,I/ Ap.

Some of the most important potato viruses are transmitted by aphids, especially the green peach aphid *(Myzus persicae)*. Others are transmitted by leafhoppers, nematodes, fungi, and sap. If a virus does not require an incubation period in the body of the insect it is called stylet-borne. If an incubation period is required, it is referred to as circulative. If the virus multiplies in the body of the insect, it is characterized as propagative.

II. POTATO VIRUS A

Potato virus A, Murphy and McKay (PVA) is also known as potato mild mosaic virus, Holmes; potato supermild mosaic, Quanjer; and potato common mosaic, Quanjer (Miller and Pollard, 1977; Smith, 1972). The disease caused by potato virus A is commonly referred to as potato common mosaic or potato mild mosaic. However, some investigators consider mild mosaic to be caused by a combined or synergistic action of PVA and PVX. French names for this disease are *mosaique benigne de la pomme de terre* or *mosaique voilee*. Spanish-speaking people call it *mosaico comun do la papa*. It is referred to as *Krauselkrankheit, Rauhmosaik der Kartoffel, A-Virus Mosaik,* or *leichtes Kartoffelmosaik* in German (Miller and Pollard, 1977). Apparently PVA was partially responsible for degeneration or "running out" of potatoes along with potato leafroll virus (PLRV). Wortley (1915) demonstrated tuber transmissibility prior to 1915. PVA is common and widespread in Europe and North America. It occurs in most other potato-growing areas also where it has been introduced through infected seed potatoes. Lee (1976) identified three strains of PVA in Taiwan.

PVA virus particles are long flexuous rods, 730 × 11 nm. They resemble PVY particles in size and structure and are indistinguishable from them. The cryptogram for PVA is */*:*/*:E/E:S/Ap.

A. Symptoms

As the various names imply, symptoms of mild mosaic are rarely very severe. Foliar symptoms usually consist of a mild mottling and slight crinkling or ruffling of infected leaves (Fig. 4.1). These symptoms are sometimes masked and difficult to detect in bright sunlight, making it almost impossible to rogue out diseased plants completely and effectively. Tuber symptoms are usually unrecognizable, although a slight reduction in size may occur. Some cultivars, especially European cultivars, such as British Queen, Up-to-Date, Kerr's Pink, Great Scot, International Kidney, Rhod-

Fig. 4.1. Potato leaf showing symptoms of mild mosaic. (Photo courtesy of Maine Life Sciences and Agricultural Experiment Station.)

erick Dhu, Epicure, and Sharpe's Express, are hypersensitive to PVA. Symptoms first appear as a yellow blotchy mottle on terminal leaves. Infected plants develop top necrosis and eventually die. Tubers from these plants are usually nonviable. Thus top necrotic cultivars possess a form of field immunity (Harriman, 1971; O'Brien and Rich, 1976; Rich, 1977; Smith, 1972).

B. Host Range

Potato is the only common natural host of PVA. However, many solanaceous plants have been experimentally infected (Kohler, 1960), including *Datura* spp., *Lycopersicon* spp., *Nicotiana* spp., *Solanum* spp., *Lycium* spp., and *Nicandra physaloides* (Harrison, 1971; Smith, 1972; Thornberry, 1966). *Solanum demissum* produces small bluish-black local lesions when inoculated with PVA (Kohler, 1948; Webb and Buck, 1955). Thompson (1959) used *Lycopersicon pimpinellifolium* as a test plant. Kohler's A6 *Solanum demissum* × Aguila hybrid A6 is a good local lesion indicator for detecting PVA when PVX is present (O'Brien and Rich, 1976; Raymer and Milbrath, 1957). Several strains of PVA can be detected on *Nicandra physaloides* (MacLachlan *et al.,* 1953). Singh *et al.,* (1977) used detached leaves of *Physalis floridana* to detect PVA. They were maintained in water at room temperature and were illuminated continuously at 200–300 fc. Local lesions appeared in 6–9 days.

C. Transmission

Potato virus A is readily tuber-perpetuated and is also transmitted by sap and aphid vectors. It is stylet-borne. Although *Myzus persicae* is the most common aphid vector, other aphid vectors include *Aphis nasturtii, Aphis rhamni, Macrosiphum euphorbiae, Brachycaudus helichrysi,* and possibly *Myzus circumflexus* (Harrison, 1971; Smith, 1972).

Nienhaus (1960) described an improved method of direct transmission from tubers to test plants. He cut away the cortical tissue which contains an inhibitor and exposed the vascular bundles. He also used the same technique for potato virus X (PVX) and potato virus Y (PVY).

D. Serology

Singh and McDonald (1981) succeeded in purifying PVA. It can be detected in potato by enzyme-linked immunosorbent assay (ELISA). Bawden and Sheffield (1944) considered PVA to be unrelated to PVX or PVY, but Harrison (1971) suggests there is evidence that it is serologically related to PVY.

E. Control

Mild mosaic can be controlled more or less satisfactorily by planting seed potatoes free from PVA, roguing out diseased plants as early as they can be detected, controlling aphid vectors, and killing the potato vines as early as practical. A number of cultivars, such as Cherokee, Chippewa, Earlaine, Houma, Katahdin, Kennebec, Mohawk, Russet Sebago, Sebago, and Spaulding Rose, possess field resistance. Irish Cobbler, like several European cultivars mentioned earlier, is top necrotic and thus field immune (Darling, 1959, 1977; Folsom *et al.,* 1955; MacLachlan *et al.,* 1953; Stevenson, 1949).

Newer cultivars reported to be resistant include: Abnaki (Akeley *et al.,* 1971, Alamo (Akeley *et al.,* 1968), Chieftain (Weigle *et al.,* 1968), DeSoto (Darling, 1977), Hunter (Davies *et al.,* 1963), LaSalle (Darling, 1977), Ona (Akeley *et al.,* 1962b), Onaway (Wheeler and Akeley, 1961), Pennchip (Mills, 1964), Penobscot (Simpson and Akeley, 1964), Raritan (Campbell and Young, 1970), Redskin (Darling, 1977), Seminole (Stevenson *et al.,* 1970), Shoshoni (Darling, 1977), Tawa (Darling, 1977), Wauseon (Cunningham *et al.,* 1968), Wyred (Riedl, 1968), and Yukon Gold (Murphy *et al.,* 1982).

Morel *et al.* (1968) produced clones free from PVA and PVY by culturing apical meristems in a special nutrient medium containing gibberellic acid and a high concentration of potassium. It was more difficult to eliminate PVY and PVS from infected clones. Once a clone is free from virus, special precautions must be taken to avoid reinfection.

III. POTATO VIRUS M

Potato virus M (PVM) and the disease caused by it have numerous synonyms; considerable confusion exists concerning their exact relationship and proper nomenclature. Miller and Pollard (1977) list potato paracrinkle virus as a synonym. Harrison (1971) and Smith (1972) refer to the virus as potato paracrinkle virus and list PVM as a synonym. Other synonyms are potato virus E, potato virus K, potato interveinal mosaic virus, and potato leafrolling mosaic virus (Harrison, 1971; Smith, 1972). Carnation latent virus (CLV) is also closely related (Bagnall *et al.,* 1959; Kassanis, 1961).

English names and probable synonyms for the potato disease caused by PVM are potato paracrinkle, potato interveinal mosaic, potato leafrolling mosaic and supermild mosaic (Bagnall *et al.,* 1956, 1959; Harrison, 1971; Kassanis, 1960, 1961; Smith, 1972).

French names include *mosaique internervale de la pomme de terre* and virus M. Spanish-speaking people call it *mosaico enrulado de la hoja, mosaico risada,* or *paracrinkle.* German terms for it are *Rollmosaik der Kartoffel* and *Paracrinkle - Mosaik* (Miller and Pollard, 1977).

Potato virus M virus particles are long, slightly flexuous, rods, 650 × 12 nm. The cryptogram is */*:*:*/*:E/E:S/* (Harrison, 1971).

A. Symptoms

Symptoms vary considerably depending upon the cultivar, the virus strain, other viruses involved, such as PVS and PVX, and growing conditions. Some cultivars are almost symptomless, while others develop chlorotic blotches, and a slight ruffling or waving of leaf margins. An interveinal mosaic is often manifest accompanied by clearing of the veins. Necrotic spots may develop on the upper leaf surface, accompanied by brown streaks on the veins of the under surface, petioles, and stems, similar to current-season symptoms of PVY (Miller and Pollard, 1977; Harrison, 1971). Lee (1972) identified three strains of PVM in Taiwan: severe, ordinary, and mild.

B. Host Range

In addition to potato, many other plants have been reported as experimental or diagnostic hosts. *Datura metel* and *D. stramonium* are suggested as diagnostic hosts by Smith (1972). Harrison (1971) includes *Nicotiana debneyi, Chenopodium quinoa, Phaseolus vulgaris* (Red Kidney), and *Vigna sinensis* as local lesion hosts. Other hosts listed for PVM or paracrinkle virus include *Beta vulgaris, Lycopersicon esculentum,* and several others (Thornberry, 1966). Ross (1968) found *L. chilense* to be a good indicator host. Kowalska and Waś (1976) recommend *L. chilense* as the most reliable test plant for PVM, although red kidney bean reacted more quickly (4–8 days) in contrast to 12–24 days for *L. chilense.*

C. Transmission

Potato virus M can be transmitted mechanically, especially if an abrasive such as carborundum is used. Some strains can also be transmitted by the aphid *Myzus persicae.* Apparently some strains of the virus can be transmitted more readily than others (Smith, 1972). Wetter and Volk (1960) succeeded in transmitting four isolates of PVM with *M. persicae* after 48 to 72 hr feeding time. Apparently PVM is circulative and not stylet-borne. They failed in their attempts to transmit their paracrinkle isolate. The virus is tuber-perpetuated and spreads from one place to another in this fashion.

D. Serology

High titer antisera of this virus can be prepared. Its precipitation end point is about $\frac{1}{64}$. Although the gel diffusion test is unsatisfactory, the chloroplast agglutination test works well. PVM is closely related to or is synonymous with potato paracrinkle virus. It is also related to carnation latent virus and PVS. However, PVM and PVS are not cross-protectant (Bagnall *et al.,* 1959; Kassanis, 1960).

E. Control

This disease can be controlled by planting virus-free seed potatoes, by controlling aphids, especially *M. persicae,* and by sanitation to reduce the risk of mechanical spread. Virus-free seed lots can be produced by apical stem culture (O'Brien and Rich, 1976).

IV. POTATO VIRUS S

Potato virus S (PVS) was not discovered until the early 1950s. Probably the reasons are that it usually produces few, if any, symptoms and that it is related to other potato viruses.

English-speaking people refer to the virus simply as potato virus S or PVS. French names are *virose masquée de la pomme de terre* and *maladie latente. Virus S de la papa* is the Spanish designation for it, and in German it is called *S-Viruskrankheit der Kartoffel, Kartoffel-S-Virus,* or *leichtes Kartoffelmosaik* (Miller and Pollard, 1977).

Potato virus S virus particles are long, slightly flexuous rods. Brandes *et al.* (1959b) recorded their size as 657×12–13 nm, while Wetter and Brandes (1956) give their dimensions as 652×13 nm. Harrison (1971) records their size as 650×12 nm and DeBokx (1969) claimed that they varied from 641 to 655 nm. Thus, they are indistinguishable from PVM (including paracrinkle and CLV) by electron microscopy. The cryptogram for PVS is */*:*/*:E/E:S/*.

A. Symptoms

Because the symptoms of PVS on most cultivars are lacking, very mild, or inconspicuous, PVS went unrecognized for several years and was not distinguished from PVX or normal effects of maturity. Under ideal conditions for symptom expression some cultivars exhibit a mild rugosity and a more open growth habit. A limpness or wilting may manifest itself on some of the older infected plants. Certain other cultivars may develop a mild mosaic, and leaves of infected plants become bronze with tiny necrotic spots on the abaxial surface. The only tuber symptom is a reduction in size which may result in a 10–20% yield reduction (Bagnall and Larson, 1957; Harrison, 1971; Miller and Pollard, 1977; Smith, 1972).

B. Host Range

Potato is the only known natural host. Indicator plants include *Chenopodium quinoa* and detached leaves of *Solanum demissum*. *Solanum demissum* leaves required less time but were less accurate than *C. quinoa* (Kowalska and Waś, 1976). Ross (1968) recommended the use of *Lycopersicon chilense* by rubbing the leaves with the cut surface of infected tubers.

Bagnall and Larson (1957) listed *Chenopodium album, Cyamopsis tetragonoloba, Nicotiana debneyi, Saracha umbellata,* and *Solanum rostratum* as hosts which produced characteristic symptoms. They also listed *Physalis philadelphica, Datura metel,* and *Solanum villosum* as symptomless hosts. Kowalska (1977) found that PVS produces minute local lesions on *Phaseolus vulgaris* cv. Red Kidney 5 days after inoculation. Książek (1975) recorded *Chenopodium album* as a weed host.

C. Transmission

Potato virus S is transmitted easily by mechanical inoculation with infectious sap. Undoubtedly it spreads naturally in the field by leaf contact. Animals, machinery, and man may also spread it from plant to plant in the field. No aphid vectors are known. It is tuber-perpetuated and spread from one location to another in infected seed potatoes (Harrison, 1971; O'Brien and Rich, 1976). It is not transmitted through true seed (Goth and Webb, 1975).

D. Serology

Potato virus S is serologically active and was first discovered by serological methods. Agglutination tests are preferable to gel diffusion tests, possibly due to the large particle size. The precipitation end point is about $\frac{1}{64}$ (Harrison, 1971). It is serologically related to other viruses of similar size such as PVM, potato paracrinkle virus, and carnation latent virus (Kassanis, 1956). Polyethylene glycol is used to aid in precipitation. Bagnall and Larson (1957) recognize strains of PVS, but Harrison (1971) does not.

E. Control

Use of virus-free seed potatoes is the best method of controlling this disease. Canadian plant pathologists have developed virus-free clones of several cultivars by means of meristem culture (see PVX, Section V). Sanitation and approved cultural practices must be used to maintain these clones relatively free from PVS. Seed lots should be tested periodically by use of serology or indicator plants to ascertain their freedom from PVS.

Saco is considered to be field resistant by some workers and immune by others (Alfieri and Stouffer, 1957; Baerecke, 1967; Bagnall and Young, 1959). Loshitskii, Kandidat, and Ogonek are resistant to PVS in Belorussia (Ambrosov and Sokolova, 1976).

V. POTATO VIRUS X

Potato virus X, Smith, is also known as potato latent virus, Burnett and Jones (Miller and Pollard, 1977). Other synonyms include potato mottle virus, potato interveinal mosaic virus, tobacco ringspot virus, Up-to-Date streak virus, potato virus B, potato virus D, healthy potato virus, and potato mild mosaic virus (Harrison, 1971; Smith 1972). At present it is called potato virus X and referred to as PVX. Its geographic range is world wide.

English-speaking people refer to the disease caused by PVX as potato latent mosaic or potato mottle. The French call it *virose masquée de la pomme de terre* or *mosaique légère*. Spanish names for the disease are *mosaico leve* and *mosaico latente de la papa*. Germans refer to it as *Kräuselmosaik* or *leichtes Mosaik der Kartoffel* (Miller and Pollard, 1977).

The virus particle is a flexuous rod approximately 515×12 nm. Its cryptogram is R/1:2.1/6:E/E:S/(Fu).

A. Symptoms

Symptoms on potato vary considerably with the cultivar and with the strain of the virus. Many of our older cultivars were 100% infected but plants remained symptomless, hence the name healthy potato virus. At one time, it was thought that the virus did not harm these symptomless cultivars, but careful experimentation showed that it reduced yields from 10 to 25%, depending on the cultivar and the strain of the virus (Harrison, 1971; Hoyman, 1964; Munro, 1961; Murphy *et al.,* 1966; Rich, 1977; Teri *et al.,* 1977; Wright, 1970, 1977). Under cool, cloudy conditions these normally symptomless cultivars may develop a mild mottling of the leaves, thus accounting for names such as mottle, interveinal mosaic, mild mosaic, and supermild mosaic. This symptom has also been referred to as weather mottling, especially by seed growers and inspectors who recognized that it was associated with a prolonged period of dark, cloudy weather. It disappears or is masked by several days of bright sunlight.

Some of the older European cultivars (Arran Crest, Epicure, Great Scot, King Edward) develop a top necrosis and usually die following infection with PVX (Rich, 1977; Smith, 1972). Tubers of these cultivars may also exhibit a corky necrosis (Smith, 1972). Cariboo and York are two new Canadian cultivars which have the same characteristic (Johnston *et al.,* 1970; Maurer *et al.,* 1968; Rich, 1977). Therefore, plants of these cultivars which appear to be healthy are PVX-free. If the top necrosis is sufficiently severe the virus is self-limiting.

B. Host Range

The natural host range of PVX appears to be limited almost entirely to potato and tomato. It causes double virus streak of tomato in combination with tobacco mosaic virus (TMV) (Smith, 1972).

Experimental hosts include about 15 angiosperm families. Plants which can become infected with PVX include *Nicotiana tabacum, N. glutinosa, Datura stramonium, Solanum dulcamara, S. nigrum, Hyoscyamus niger, Cyphomandra betacea, Petunia* sp., crimson clover *(Trifolium incarnatum),* and other legumes. Symptoms vary with the strain of PVX and the species or cultivar of the host plant (Fribourg and de Zoeten, 1975; Harrison, 1971; Smith, 1972; Willis and Larson, 1960).

Diagnostic hosts include *Datura stramonium, Gomphrena globosa,* and *Chenopodium amaranticolor.* PVX causes a systematic mottle on *D. stramonium.* It produces distinctive necrotic local lesions on *G. globosa* (Fig. 4.2) in 4–5 days, and chlorotic, necrotic, or ring spot local lesions on *C. amaranticolor* (Harrison, 1971). The author prefers *G. globosa.* It is easy to grow, produces distinctive lesions, and infection can be quantified by counting the number of lesions. However, Francki (1967) warns that excessively high light intensity can produce spontaneous lesions resembling those produced by PVX. Mendoza and Haynes (1974) used detached *G. globosa* leaves in petri plates under continuous fluorescent light for early detection of PVX. Singh (1972) used spinach *(Spinacia oleracea)* as a test plant for PVX. Merriam and Akeley (1974) found potato seedling CC31–4a to be a good local lesion test plant. It may be necessary to use other diagnostic hosts to distinguish between strains of PVX.

C. Transmission

This virus is transmitted easily by infectious sap, either from tubers or leaves. It is also transmitted by the cutting knife, mechanical planter, cultivating and spraying equipment, animals, and by contact of sprouts, leaves, or roots (Mai, 1947; Manzer and Merriam, 1961; Roberts, 1946). Transmission by grasshoppers has also been reported. There is evidence that it can be transmitted by spores of the potato wart disease fungus, *Synchy-*

Fig. 4.2. *Gomphrena globosa* plant with one leaf inoculated with PVX. Note local lesion symptoms.

trium endobioticum. It is not transmitted by true seeds but can be transmitted by dodder, *Cuscuta campestris* (Harrison, 1971; Rich, 1977; Smith, 1972). Wright (1974) discovered that PVX retained its infectivity for 3 hr on unpainted wood, iron, rubber, and human skin; 6 hr on painted wood, jute, and cotton; and 24 hr on soil.

D. Serology

Potato virus X is serologically active. Antisera can be prepared by injecting rabbits intramuscularly with heat-clarified sap plus adjuvant. Precipitin tests can be conducted in test tubes, or chloroplast agglutination tests, using crude sap, can be performed by mixing drops of sap and antiserum on slides. The gel diffusion test is not recommended. Wright *et al.*

(1977) prefer the latex agglutination test, which is reliable, fast, and very sensitive. The latex increases visibility of clumping. This test is used routinely in maintaining certified seed as nearly free from PVX and PVS as possible.

Virus in sap has a precipitation end point of up to 1/5000. The thermal inactivation point of PVX in sap is about 70°C. The dilution end point is $1:10^{-5}$ to $1:10^{-6}$. Longevity *in vitro* at room temperature can vary from several weeks to a year, depending on the strain of the virus (Harrison, 1971; Smith, 1972).

E. Control

Latent mosaic or mottle, caused by PVX, can be controlled initially by producing new cultivars from true seed. However, it is difficult, if not impossible, to maintain stocks free from PVX. Sanitation is important. Growers should plant virus-free seed potatoes if possible. Seed stock can be tested on *Gomphrena globosa* or by using the chloroplast agglutination test. PVX has a tendency to build up rapidly in seed stocks if not checked regularly (Manzer *et al.,* 1975). Plants should be cultivated when they are small. Herbicides may reduce the number of cultivations required. Use of aerial spray equipment (airplanes or helicopters) reduces the possibility of sap transmission by contaminated equipment (Rich, 1977).

Until recently, old cultivars, such as Green Mountain and Russet Burbank, were 100% infected with PVX. This virus has been eliminated from certain clones of these cultivars by apical stem culture and/or heat (Kassanis and Varma, 1967; MacDonald, 1973; Rich, 1969; Stace-Smith and Mellor, 1968; Wright and Cole, 1976).

Potato seedling S41956, Saco, Saphir, and Tawa are immune to PVX (Akeley *et al.,* 1955b; Peterson and Hooker, 1958; Wetter, 1961). Additional resistant cultivars include Hunter (Davies *et al.,* 1963), Raritan (Campbell and Young, 1970), Reliance (Hoyman *et al.,* 1963), and Wauseon (Cunningham *et al.,* 1968). Murphy *et al.* (1982) also list Butte, Russette, and Yukon Gold as resistant cultivars. They indicate that C7358–26A, C74109–8, CAO2–7, CC26–1A, and W564–3A are resistant seedlings or clones.

VI. POTATO VIRUS Y

Potato virus Y, Smith, is also known as potato vein-banding virus, Johnson; streak virus, Orton; leaf-drop streak virus, Murphy; stipple streak virus, Atanasoff; acropetal necrosis virus, Quanjer; vein-banding virus,

Valleau and Johnson; and potato severe mosaic virus, Samuel (Miller and Pollard, 1977; Smith, 1972). Harrison (1971) lists tobacco veinal necrosis virus and potato virus C as strains of PVY. Darby *et al.,* (1951) also recognize several strains of PVY. The disease caused by PVY is often referred to as vein-banding, leaf-drop streak, severe mosaic, or rugose mosaic (Miller and Pollard, 1977). At one time rugose mosaic was considered to be caused by the combined or synergistic action of PVX and PVY (Koch, 1933); some investigators still make this distinction (Miller and Pollard, 1977).

French names for the disease are *bigarrure de la pomme de terre* and *bigarrure-affaisement.* Spanish-speaking people call it *mosaico severo, mosaico suave,* and *necrosis de nervaduras en papa.* The Germans refer to it as *Strichelkrankheit der Kartoffel* or *Y-virus Mosaik* (Miller and Pollard, 1977).

The virus particle is a long, flexuous rod. It measures about 730 × 11 nm and is indistinguishable from PVA in size and shape. Its cryptogram is */*:*/*:E/E:S/Ap.

A. Symptoms

Chronic symptoms consist of chlorotic mottling, severe rugose wrinkling, moderate to severe dwarfing, and premature death (Figs. 4.3 and 4.4). Tubers are smaller than normal. Current-season symptoms are characterized by the development of brown necrotic streaks along the veins of the leaves, the petioles, and the stems. In severe cases, the leaves and petioles die and hang downward from the stems, hence the name "leaf-drop streak." Severity of symptom expression varies considerably from one cultivar to another. If infection takes place late in the season, tubers may become infected without development of plant symptoms (Harrison, 1971; O'Brien and Rich, 1976; Rich, 1977). Smith (1972) described the differences in disease reaction by a number of different cultivars. Schultz *et al.* (1947) described a variety of symptoms varying from light green, mottled, and rugose leaves to severely necrotic, curled, and dwarfed plants, depending on the cultivars involved.

B. Host Range

Potato virus Y affects about eight different families of angiosperms. *Nicotiana tabacum, N. glutinosa, Solanum dulcamara, S. nigrum, Lycopersicon esculentum, Cyphomandra betacea, Hyoscyamus niger, Petunia* spp., and *Chenopodium amaranticolor* are some of the more common hosts. It is systemic with vein clearing and vein banding in *N. tabacum. Chenopo-*

Fig. 4.3.   Potato plant on left showing rugose mosaic symptoms. Plant on right is healthy. (Photo courtesy of Maine Life Sciences and Agricultural Experiment Station.)

dium amaranticolor is a local lesion indicator (Harrison, 1971; Smith, 1972). Other hosts include *Capsicum frutescens, Dahlia* spp., *Physalis* spp., and many others (Thornberry, 1966). Ross (1948) listed several others but favored the use of *Physalis floridana* as a diagnostic test plant. Tomatoes suffered economic losses from PVY in Florida (Cox, 1965). Edwardson (1974a,b) published two monographs on the PVY group.

C. Transmission

Potato virus Y is mechanically transmitted with comparative ease. It is tuber-perpetuated but is not transmitted by true seed. Many species of aphids are capable of transmitting the virus. It is stylet-borne, as no incubation period is required. The green peach aphid *(Myzus persicae)* is the most common and efficient aphid vector. Other aphid vectors include *M. certus, M. ornatus, Aphis rhamni, A. nasturtii, Brachycaudus helichrysi,* and *Macrosiphum euphorbiae* (Harrison, 1971; Smith, 1972). Shands and

Fig. 4.4.   Close-up of potato leaves showing rugose mosaic.

Simpson (1971) found *A. nasturtii* to be an important vector of PVY in Maine. Schultz (1963) also indicated that PVY is transmitted by a mite *Tetranychus telarius).*

D. Serology

High titer antisera have been prepared, using clarified tobacco sap. The precipitation end point is about $\frac{1}{32}$. The gel diffusion test is not recommended.

Several strains of PVY react somewhat differently in serological tests.

Serological relationships also exist with other members of the potato virus Y group (Harrison, 1971; Matthews, 1970; Smith, 1972).

E. Control

High-quality seed potatoes, free from PVY should be planted, preferably in fields isolated from commercial potato fields. Diseased plants should be rogued out as early in the season as they can be detected. Aphids should be controlled with systemic insecticides, contact insecticides, or both.

Resistant cultivars should be used whenever adapted to local conditions. Unfortunately no commercial cultivars are immune to PVY. However, a number of them are reported to be field-resistant, including Avon, Katahdin, Kennebec, Monona, Nordak, Norgleam, Oromonte, Saco, Snowflake, and York (Darling, 1959, 1977; Davies *et al.*, 1975; Easton *et al.*, 1958; Folsom *et al.*, 1955; Johansen, 1963; Johnston *et al.*, 1970; Rich, 1977; Twomey *et al.*, 1968). Schultz *et al.* (1947) found that resistance is affected by aphid dosage. A recent report by Murphy *et al.* (1982) characterizes Caribe and Russette as resistant cultivars.

VII. POTATO AUCUBA MOSAIC VIRUS

Potato aucuba mosaic is caused by the potato aucuba mosaic virus (PAMV). The virus and/or the disease have several synonyms, including potato virus F, potato virus G, potato tuber blotch virus, and potato yellow mosaic virus (Clinch *et al.*, 1936; Harrison, 1971; Smith, 1972).

Potato aucuba mosaic is not a serious disease in the United States or Canada. It was formerly widespread in Great Britain and other European countries, but its importance has declined in recent years.

Potato aucuba mosaic virus particles are long, flexuous rods about 580 × 13 nm. The cryptogram for this virus is */*:*/:*:E/E:S/Ap (Harrison, 1971).

A. Symptoms

A bright yellow spotting of the lower and middle leaves characterizes the symptoms of PAMV on some cultivars, e.g., Arran Banner, Arran Crest, and Epicure. Smith (1972) suggests using Irish Chieftain as a differential host because of the bright yellow mottling all over the plant. In addition to foliar spotting, a number of cultivars, including Champion, Dunbar Yeoman, Early Regent, Great Scott, Majestic, and President, also develop tuber necrosis. Arran Victory produces large, yellow, blister-like areas on the lower leaves but no tuber symptoms are apparent. British Queen plants infected with PAMV exhibit a wilting and desiccation of the tips of the lower leaves, followed by a severe yellow mottling of the leaves and necrosis of the tubers. Irish Chieftain is characterized by a bright yellow foliar mottle, including the top leaves. This might be confused with potato calico, caused by alfalfa mosaic virus, in areas where both diseases occur, but it is usually more brilliant than are calico symptoms (Harrison, 1971; Smith, 1972).

B. Host Range

Hosts other than potato include tobacco *(Nicotiana tabacum),* tomato *(Lycopersicon esculentum),* pepper *Capsicum annuum), Chenopodium amaranticolor* (Fig. 4.5 and *Ch. quinoa* (Harrison, 1971; Juo and Rich,

Fig. 4.5. *Chenopodium amaranticolor* leaf showing local lesion symptoms after inoculation with potato aucuba mosaic virus.

1969; Smith, 1972). Potato aucuba mosaic virus causes a severe top necrosis on pepper. It causes local lesions on *Ch. amaranticolor* and *Ch. quinoa* (Juo and Rich, 1969).

C. Transmission

The virus can be transmitted mechanically by infectious sap. It is also transmitted by aphids, primarily *Myzus persicae,* but a helper virus, such as PVA or PVY, is required for aphid transmission (Harrison, 1971; Smith, 1972). Potato aucuba mosaic virus is tuber-perpetuated, and is spread from one locale to another in this manner.

D. Serology

A PAMV antiserum has been developed. The virus is unrelated to PVA or PVY even though they serve as helper viruses in aphid transmission. There are many different strains of PAMV (Harrison, 1971; Smith, 1972).

E. Control

This disease has been declining in recent years. New cultivars free from the disease have been developed. Even though they may not be resistant, it is relatively easy to keep these cultivars free from PAMV. Disease-free seed should be planted. Aphid control should be practiced, and diseased plants should be rogued out if they occur.

VIII. ANDEAN POTATO LATENT VIRUS

Andean potato latent virus (APLV) was first described by Gibbs *et al.* (1966). It occurs in several of the Andean countries of South America, including Bolivia, Colombia, and Peru (Fribourg *et al.,* 1977). It has a wide host range among the Solanaceae, Amaranthaceae, and Chenopodiaceae.

A. Symptoms

As the name implies, primary symptoms are usually latent. However, secondary symptoms often include chlorotic minor vein netting and/or mild mosaic and rugosity (International Potato Center, 1977). The effect on tuber size and yield is not known, but probably they are less than normal.

B. Causal Agent

The causal agent of this disease is Andean potato latent virus. Apparently several strains of the virus occur in nature (Fribourg *et al.,* 1977). According to Gibbs and Harrison (1969), APLV is a strain of eggplant mosaic virus (EMV) and belongs to the tymovirus group.

C. Transmission

Andean potato latent virus is readily transmitted from diseased to healthy plants by brushing of the leaves against each other; therefore, it is sap-transmissible. It is also tuber-perpetuated. Jones and Fribourg (1977) found that a species of flea bettle *(Epitrix* sp.) could transmit APLV, but its efficiency was low. They also reported a low level of transmission through the true seed.

D. Control

Roguing is not very effective due to the latent nature of APLV. It can be controlled by selecting and propagating clones which are known to be free of the virus (International Potato Center, 1977). Clones can be tested for presence or absence of APLV serologically or by using *Nicotiana bigelovii* Wats. as an indicator plant (Fribourg *et al.,* 1977). Control of flea bettles may be helpful also.

IX. POTATO LEAFROLL VIRUS

Potato leafroll virus, Appel (PLRV) is also known as potato phloem necrosis virus, Quanjer (Smith, 1972). English names for the disease caused by PLRV are potato leafroll, potato leaf curl, and potato phloem necrosis or net necrosis. French-speaking people call the disease *enroulement de la pomme de terre* or *feuilles en cuillère.* Spanish names are *enrollado de las bojas* and *arrollamiento de las bojas de la patata.* Germans refer to the virus as *Blattrollvirus* and the disease as *Blattrollkrankheit der Kartoffel* (Miller and Pollard, 1977).

This is one of the oldest and most serious of the virus diseases of potato. It was probably largely responsible for the "running out" of old potato cultivars before it was known that viruses were responsible for their degeneration. It not only causes a marked reduction in yield, but it also causes a tuber necrosis, often called net necrosis or phloem necrosis in cultivars which are susceptible to this phase of the disease.

Potato leafroll virus is small, isometric, and measures about 24–25 nm in diameter (Smith, 1972). Its cryptogram is */*:*/*:S/S,I/Ap.

Webb *et al.* (1952) recognized four strains of PLRV, based primarily on symptoms produced on *Physalis floridana* at different temperatures. Rich (1951) compared leafroll in the western and eastern United States and concluded that they were comparable.

A. Symptoms

Plants grown from infected tubers are pale, dwarfed, more upright than normal, and, as the name implies, the leaves are rolled, especially the lower ones (Fig. 4.6). They are thick, leathery, brittle in texture, and crackle when squeezed in the hand. Symptoms first appear about a month after planting or when the plants are about 6 in. high. Diseased plants produce fewer and smaller tubers than normal plants, resulting in a marked yield loss.

Plants which become infected during the growing season (current or primary infection) exhibit little if any dwarfing. Early season infection usually results in a characteristic rolling of the upper leaflets (Fig. 4.7). A purple pigment at the base of young leaflets may develop. Plants infected late in the growing season usually remain symptomless. However, tubers produced by these plants often develop phloem necrosis (Fig. 4.8) (net necrosis) unless they are immune to this phase of the leafroll syndrome (O'Brien and Rich, 1976; Rich, 1977; Webb and Schultz, 1958b). The Green Mountain cultivar grown in northern New England does not exhibit phloem necrosis at harvest time but it frequently develops during storage in tubers produced by plants infected with primary leafroll (Folsom and Rich, 1940; Gilbert, 1928; Rich, 1951). This is an important factor in the loss of popularity of this once important cultivar. Rich (1951) found that Russet Burbank tubers grown in the state of Washington from plants infected with primary leafroll often exhibit phloem necrosis at harvest time and that it develops further during storage. However, necrosis was not observed in tubers from chronically infected leafroll plants. More recently, however, Guthrie (1959a, 1961) reported the occurrence of this phenomenon in Russet Burbank tubers grown in Idaho. Green Mountains grown in Rhode Island and Long Island, New York, do not develop phloem necrosis even though 60–75% of the plants become infected with current-season leafroll (A. E. Rich, personal observation). This is probably due to the high soil temperature during the growing season. Recently Manzer *et al.*, (1982) reported that stem-end browning of potato tubers is associated with early season infection with PLRV.

Another tuber symptom of leafroll is spindling sprout. Tubers infected with leafroll, especially those which also exhibit phloem necrosis, often develop long, spindly sprouts (Fig. 4.9). However, this symptom is not 100%

Fig. 4.6.   Potato plant infected with potato leafroll virus.

accurate as a diagnostic character because other diseases occasionally produce spindling sprouts.

Gibson (1974, 1975) describes a disease which he calls top-roll. It is not tuber transmitted but is associated with feeding by the aphid *Macrosiphum euphorbiae*.

B. Host Range

In nature, although potato is the principal host of PLRV, the virus occasionally also attacks tomato *(Lycopersicon esculentum)*. Other solanaceous hosts include *Solanum dulcamara, S. villosum, Datura stramonium,*

Fig. 4.7.   Potato plant showing current-season leafroll symptoms.

D. tatula, Physalis angulata, and *P. floridana. Physalis floridana* is used widely as an indicator plant. Symptoms consist of dwarfing, chlorosis, leaf-rolling, and phloem necrosis (Smith, 1972). Nonsolanaceous host plants include *Amaranthus caudatus, A. graecizans, A. retroflexus, Celosia argentea, Gomphrena globosa,* and *Nolana lanceolata* (Natti *et al.,* 1953). Thornberry (1966) lists several other possible hosts.

C. Transmission

Potato leafroll virus, like most other potato viruses, is tuber-perpetuated. It is not sap-transmissible, however. It is transmitted from plant to plant

Fig. 4.8. Potato tuber showing phloem necrosis following current-season infection with the leafroll virus. (Photo courtesy of Maine Life Sciences and Agricultural Experiment Station.)

by aphids, primarily the green peach aphid *(Myzus persicae)*. Other possible aphid vectors include the buckthorn aphid *(Aphis nasturtii),* the potato aphid *(Macrosiphum euphorbiae),* and the foxglove aphid *(Acyrthosiphon solani)* (Folsom *et al.,* 1955; Shands *et al.,* 1972; Simpson, 1977; Simpson and Shands, 1949). The virus is persistent in *M. persicae* and requires an incubation period. This is probably true of all other aphid vectors as well. Shands *et al.* (1972) found *A. nasturtii* to be a poor vector. The virus can also be transmitted experimentally by grafting.

D. Serology

A PLRV antiserum has been developed, thus permitting the enzyme-linked immunosorbent assey (ELISA) to be used to detect the virus (Clarke, 1981; Maat and Bokx, 1978). Duffus and Gold (1969) compared PLRV serologically with beet western yellows virus by membrane feeding of aphids. They concluded that the two viruses were not closely related, even though they produced similar symptoms on *P. floridana* and *D. stramonium*. More recently Duffus (1981a,b) reported that beet western yellows virus (BWYV) is often associated with potato leafroll.

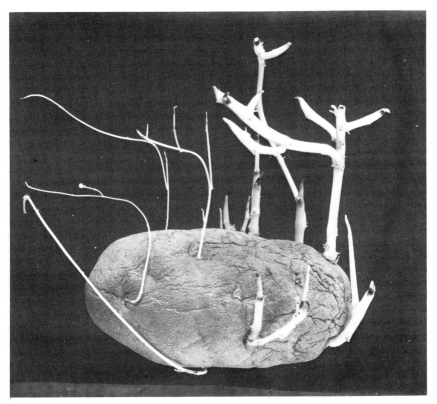

Fig. 4.9.   Spindling sprouts on stem end of tuber due to leafroll and phloem necrosis infection. (Photo courtesy of Maine Life Sciences and Agricultural Experiment Station.)

E. Control

The most important control measure for potato leafroll is to plant virus-free seed potatoes. Other important practices include isolation from other potato fields, control of aphid vectors, roguing out diseased plants to reduce sources of inoculum, and early harvesting or vine killing of seed plots and certified fields (Wright and Hughes, 1964). Aphids can be controlled by the use of systemic insecticides or potato plants can be sprayed or dusted with contact insecticides (Pond, 1964). Because aphids are sucking insects, insecticides which are primarily stomach poisons are ineffective. Eradication of the Canada plum *(Prunus nigra)* is recommended in northern Maine (Holbrook and Kleinschmidt, 1975; Shands and Simpson, 1969). Canada plum is the natural overwintering host of the green peach aphid in that region.

Folsom and Stevenson (1946) recognized that some cultivars and seedlings are more resistant than others. Resistant varieties include Abnaki (Akeley *et al.*, 1971), Belrus (Murphy *et al.*, 1982), Cascade (Hoyman, 1970), Penobscot (Simpson and Akeley, 1964), and Yukon Gold (Murphy *et al.*, 1982). Darling (1977) also lists Houma and Yampa as resistant and Katahdin as field resistant. Kennebec is also somewhat resistant (MacKinnon, 1970). Chippewa, Green Mountain, and Russet Burbank are highly susceptible to natural infection under field conditions (Locke, 1947; Rich, 1951).

Many cultivars, including most of those released in recent years, do not develop net necrosis in the tubers even though they are susceptible to leafroll. Darling (1977) lists Cherokee, Chippewa, Earlaine, Houman, Katahdin, Kennebec, Merrimac, Redskin, Russet Sebago, Saco, and Sebago as resistant to net necrosis. However, this list is not intended to be complete. Many of the older cultivars, such as Green Mountain, Irish Cobbler, and Russet Burbank, are highly susceptible to net necrosis. Storing Green Mountains at 33° or 70°F will minimize development of net necrosis in tubers infected with leafroll (Folsom *et al.*, 1955; Folsom and Rich, 1940). For further information on net necrosis readers are referred to Davidson and Sanford (1955); Folsom (1946); Folsom and Rich (1940); and Rich (1951).

Stem-end browning has often been confused with net necrosis (Folsom and Rich, 1940; O'Brien and Rich, 1976). The discoloration in stem-end browning tubers is usually darker brown, more shallow, and less netted than the discoloration in net necrosis tubers. Recently the leafroll virus has been implicated as a probable cause of stem-end browning as well as causing net necrosis (Manzer *et al.*, 1982). Fortunately, most present day cultivars are resistant to stem-end browning as well as net necrosis.

X. CORKY RING SPOT

Corky ring spot of potatoes occurs in the United States, Europe, Indonesia, and South Africa (Eddins *et al.*, 1946; O'Brien and Rich, 1976; Rich, 1977; Smith, 1972; Smith and Wilson, 1978; Walkinshaw and Larson, 1959). It is a complex disease involving two or more viruses or virus strains. The nomenclature of the disease is confusing. It is usually referred to as corky ring spot in the United States (Webb and Schultz, 1958a). Some Europeans call it Spraing, while others may refer to it as sprain or internal rust spot. It may be called *Eisenfleckigkeit* in German. Stem mottle is applied to the foliar stage of one phase of the disease (see references above, Lihnell, 1959, and Rich, 1951).

Corky ring spot is caused primarily by the tobacco rattle virus (TRV).

The potato mop-top virus produces similar symptoms (Harrison, 1971; Smith, 1972). The cryptogram for tobacco rattle virus is R/1:2.3/5:E/E:S/ Ne (Harrison, 1971).

The virus particles are straight, tubular, helically constructed, about 25 nm in diameter, and of varying length. The longest particles measure about 190 nm long and the shorter particles vary from 45 to 115 nm depending on the isolate (Harrison, 1971).

A. Symptoms

Affected potato tubers exhibit corky cracks and brown arcs or rings on the surface and in their tissues. The characteristic concentric rings and arcs aid in distinguishing it from noninfectious internal brown spot and tuber necrosis caused by other viruses, such as alfalfa mosaic, potato leafroll, and potato yellow dwarf (Smith and Wilson, 1978). Tubers may be misshapen, and severe symptoms make the tubers unsalable.

Some isolates of TRV cause stem mottle of affected potato plants and an aucuba-like symptom on affected leaves. Apparently this phase of the disease is more common in Europe that it is in the United States (Harrison, 1971).

B. Host Range

Tobacco rattle virus has a wide host range. It affects several species of *Nicotiana* including *N. tabacum* and *N. clevelandii*. Other crop hosts include aster *(Callistephus chinensis),* gladiolus *(Gladiolus hortulanus),* tulip *(Tulipa gesneriana),* hyacinth *(Hyacinthus orientalis),* sugar beet *(Beta vulgaris),* and pepper *(Capsicum frutescens).* Many weeds, especially chick weed *(Stellaria media)* serve as hosts in the field (Davis and Allen, 1975; Harrison, 1971). *Phaseolus vulgaris* and *Chenopodium amaranticolor* are diagnostic hosts which produce local lesions (Harrison, 1971).

C. Transmission

Tobacco rattle virus is sometimes but not always tuber-perpetuated. It is transmitted by grafting. It is readily transmitted naturally by several species of nematodes, including *Trichodorus pachydermus* and *T. primitivus* (Harrison, 1964, 1971).

D. Control

Disease-free seed potatoes, crop rotation, and weed control will help to keep this disease in check. Soil fumigation with nematicides will control the nematode vectors and thus control the disease (Dallimore, 1972). Use of systemic insecticides and/or nematicides, such as aldicarb, mocap, oxamyl, and phorate, will also reduce losses from corky ring spot (Alphey *et al.,* 1975; Brown and Sykes, 1973; Cooper and Thomas, 1971; Sykes, 1975). Pungo, Green Mountain, Hudson, Superior, Plymouth, Merrimack, and U.S.D.A. seedling B6969-2 are tolerant or resistant (Weingartner *et al.,* 1977).

XI. YELLOW DWARF VIRUS

Potato yellow dwarf virus (PYDV) occurs in northeastern United States and eastern Canada. The English name for the disease is yellow dwarf. French-speaking people call it *nanisme jaune de la pomme de terre* or *jaunisse.* Spanish and German translations are *nanismo amarillo* and *Gelbzwergigkeit,* respectively (Miller and Pollard, 1977).

Apparently there are four strains of PYDV. They are identified as the New York strain, New Jersey strain, strain B5, and a vectorless strain. Strain differences are based on symptoms and vectors (Smith, 1972). PYDV has large bacilliform particles, 380 × 75 nm. Its cryptogram is */*:*/*:U/*:5/ Au. (Martyn, 1968).

A. Symptoms

The most obvious symptoms of yellow dwarf are a dwarfed condition and yellow color—hence its name. The growing apex dies early in the season. Necrosis occurs in the upper portion of affected plants. Brown specks are visible in the pith and cortex. Tubers are few, small, sessile, or on short stolons. Some are irregular in shape and cracked. Internal tuber necrosis may appear as rusty brown specks or areas surrounding the pith area and in the cortex. The necrotic lesions are most prevalent at the middle and bud end of infected tubers, while the stem end usually remains symptomless (Barrus and Chupp, 1922). Tuber symptoms are easily confused with nonparasitic internal brown spot or sprain (Walker and Larson, 1939).

B. Host Range

Hosts, in addition to potato, include tomato *(Lycopersicon esculentum),* red clover *(Trifolium pratense),* and crimson clover *(T. incarnatum).* Experimentally infected plants include *Nicotiana glutinosa, N. glauca, N. rustica, N. tabacum* Turkish, *Physalis pubescens, Solanum melongena, Datura stramonium, Vicia faba,* and *Callistephus chinensis* (Black, 1937, 1938;

Smith, 1972; Walker, 1952). Crimson clover and *N. rustica* can be used to distinguish between the New York and New Jersey strain of PYDV.

C. Transmission

The New York strain of PYDV is transmitted by the clover leafhopper *Aceratagallia sanguinolenta.* The New Jersey strain of PYDV is transmitted by two species of Agallian leafhoppers, *Agallia constricta* and *A. quadripunctata.* Ishihara (1969) lists several other species of leafhoppers in the genera *Aceratagallia, Agallia, Erupoasca,* and *Agalliopsis* as vectors of PYDV. The virus is not stylet-borne but is circulative, requiring an incubation period of 6 to 10 days in the leafhoppers before they become infective. Sometimes the virus overwinters in the bodies of leafhoppers (Smith, 1972; Walker, 1952).

The virus is also sap-transmissible with some difficulty. A vectorless strain of PYDV has been identified by L. M. Black (Smith, 1972).

D. Control

Disease-free, certified seed potatoes should be planted in isolated areas away from clover fields. Diseased plants should be rogued out as early as they can be detected. Control of leafhoppers with insecticides is beneficial (Hodgson *et al.,* 1974; Rich, 1977). The Sebago cultivar is resistant (Rieman and McFarlane, 1943).

XII. ALFALFA MOSAIC VIRUS

Alfalfa mosaic virus causes a disease of potato known as potato calico (Black and Price, 1940). The disease in alfalfa is called alfalfa mosaic or lucerne mosaic in English, *mosaique de la luzerne* in French, *mosaico de la alfalfa* or *mosaico de la lucerna* in Spanish, and *Luzernemosaik* in German (Miller and Pollard, 1977). Folsom *et al.,* (1955) state that potato calico is caused by a clover mosaic in Maine.

Infected potato plants contain both rod-shaped and isometric particles. The cryptogram for the virus is R/1:1.3/18:u/u:S/Ap (Harrison, 1971).

A. Symptoms

Infected potato plants exhibit a distinct yellow mottling of leaves (Fig. 4.10). Leaf distortion and/or necrosis may or may not be manifest. Tubers may be misshapen, cracked, and few in number (Hodgson *et al.,* 1974; Oswald, 1950).

Fig. 4.10.   Potato calico caused by alfalfa mosaic virus.

B. Host Range

Alfalfa mosaic virus infects not only alfalfa, but also peas, tobacco, and potato. It may be present in clover in Maine, thus accounting for the report that a clover mosaic causes potato calico in that state (Folsom *et al.,* 1955). Very little alfalfa was grown in Maine at that time.

C. Transmission

The virus is easily transmitted to *Nicotiana tabacum* by sap inoculation. Aphids, especially *Acyrothosiphon pisi,* are important insect vectors (Harrison, 1971). It is also tuber-perpetuated (Rich, 1977).

D. Control

Only healthy seed potatoes should be planted, preferably away from alfalfa and clover fields. Volunteer potato and alfalfa plants should be destroyed. An aphid control program should be followed.

XIII. POTATO MOP-TOP VIRUS

Potato mop-top virus (PM-tV) was first described as a distinct virus causing a distinct disease of potatoes in 1966. Prior to that time it was confused with tobacco rattle virus or potato aucuba mosaic virus. Other names for the disease or its causal agent include potato concentric necrosis virus and potato yellow mottling virus. It is represented by the cryptogram R/*:*/ *:E/E:S/Fu. Its particles are straight rods 16–18 nm long (Harrison, 1971). The virus is known to occur in Europe and probably in Peru.

A. Symptoms

In some cultivars (Arran Pilot, Alpha, Ulster Sceptre) the plants are stunted and the leaves distorted, hence the name mop top. Leaves of many cultivars develop golden yellow flecks or blotches resembling aucuba mosaic. Tubers often develop arcs or concentric rings resembling corky ring spot or spraing.

B. Host Range

Potato is the only known natural host. Experimental and diagnostic hosts include *Nicotiana debneyi, N. tabacum,* and *Chenopodium amaranticolor.* The virus produces local lesions consisting of concentric necrotic lines on *C. amaranticolor.* Local lesions and systemic necrotic line patterns form on *N. debneyi* (Harrison, 1971).

C. Transmission

About 50% of infected tubers transmit the virus to their progeny. The virus is sap-transmissible, at least experimentally. The principal and unique method of transmission is by the soil fungus, *Spongospora subterranea,* the causal agent of powdery scab. This fungus attacks potato roots and tubers and acts as a vector. The virus can persist in spores of *S. subterranea* for at least a year (Harrison, 1971).

D. Control

Disease-free seed potatoes should be planted in soil which is not infested with *S. subterranea*. Seed potato fields should be rogued carefully, and crop rotation should be practiced.

XIV. TOBACCO NECROSIS VIRUS

Tobacco necrosis virus (TNV) is also known as tulip Augusta disease virus and bean stipple streak virus. The French, Spanish, and German names for the disease on tobacco are *necrose du tabac, necrosis del tabaco,* and *Tabaknekrose,* respectively. The Dutch name for the potato disease caused by this virus is ABC disease. The disease has been recorded in Europe and North America and it may occur worldwide.

The cryptogram for the virus is R/1:1.5/19:S/S:S/Fu. Its particles are isometric, about 28 nm in diameter (Harrison, 1971).

A. Symptoms

Tuber symptoms may be any one of three types: (a) dark brown lesions with noticeable reticulate cracks, (b) light brown lesions with fine reticulate cracks, and (c) blisters, which collapse to form dark, sunken lesions. Plants and progeny tubers are usually symptomless (Harrison, 1971).

B. Host Range

The roots of numerous crop and weed plants are susceptible under field conditions. Species representing about forty families of angiosperms have been infected experimentally. Local lesions which are either chloratic or necrotic are common. The virus becomes systemic in tulip Augusta and causes bean stipple streak. Diagnostic indicator hosts include *Phaseolus vulgaris, Chenopodium amaranticolor, Cucumis sativus,* and *Nicotiana tabacum* (Harrison, 1971).

C. Transmission

The virus is transmitted by the fungus *Olpidium brassicae,* the same fungus which transmits lettuce big vein virus. This is the first virus to be identified as transmissible by a fungus. The fungus does not appear to colonize potato tubers. *Lactuca sativa* roots are easily infected with TNV by *O. brassicae.*

D. Control

Susceptible potato cultivars should be avoided. Crop rotation and weed control should help to keep the disease in check.

XV. TOMATO BLACK RING VIRUS

Tomato black ring virus is known to infect potatoes in Germany and to a lesser extent in Scotland. The disease is also known as potato bouquet and the virus as potato bouquet virus. Other names for the virus are bean ring spot virus, lettuce ring spot virus, beet ring spot virus (strain), celery yellow vein virus (strain), and potato pseudo-aucuba virus (strain). It is not recognized as a disease of potatoes in North America. The cryptogram for this virus is R/*:*/38/:5:5/Ne (Harrison, 1971). The virus particles are isometric, about 30 nm in diameter.

A. Symptoms

Primary symptoms, consisting of necrotic leaf spotting, are mild or nonexistent the first year following infection. In subsequent years, leaves are slightly misshapen and cupped, and plants are somewhat stunted, giving rise to the name "bouquet." Some strains cause a yellow mottling of the leaves, hence the name "pseudo-aucuba." Tuber yields are reduced about 20–30% (Harrison, 1971).

B. Host Range

The virus infects many weed and crop plants including bean, beet, celery, turnip, strawberry, raspberry, grape, peach, and cherry. Species representing more than twenty angiosperm families can be infected experimentally. Indicator hosts include *Nicotiana tabacum, N. rustica, N. clevelandii, Chenopodium amaranticolor, C. quinoa, Phaseolus vulgaris,* and *Cucumis sativus.*

C. Transmission

The virus is tuber-perpetuated from one potato crop to the next. It is sap-transmissible to susceptible host plants.

There are two serologically distinct isolates of the virus which are transmitted by two different species of nematodes in the genus *Longidorus. Longidorus attenuatus* transmits the type form and *L. elongatus* is a vector for the beet ring spot form. Infected weed seeds serve as virus reservoirs and probably serve as a method of distribution of the virus to new locations (Harrison, 1971).

D. Control

Disease-free seed potatoes should be used for planting. Crop rotation and weed control should be beneficial. Soil fumigation with a nematicide will control the nematode vectors.

XVI. POTATO SPINDLE TUBER VIROID

Potato spindle tuber, caused by the potato spindle tuber viroid, is a serious disease of potato in the United States and Canada, but apparently is not a problem in most other countries. French names for the disease are *filosites des tubercules de la pomme de terre* and *tubercules fusiformes.* Spanish-speaking people refer to it as *tuberculo puntiagudo* or *tuberculo en forma de uso.* The disease in German is *Spindelknollenkrankheit der Kartoffel* (Miller and Pollard, 1977). Its cryptogram is */*:*/*:*/*:S/* (Martyn, 1968).

A. Symptoms

Symptoms of spindle tuber on infected plants are difficult to detect and are often overlooked by potato growers and inspectors. Infected plants are somewhat smaller and more erect than healthy plants. Leaves are a slightly darker shade of green, the angle of branching is slightly more acute than normal, and leaflets tend to be twisted (Fig. 4.11). Plant symptoms show best at high temperatures and may be masked or obscured at cool temperatures (Folsom *et al.,* 1955; Goss, 1930; O'Brien and Rich, 1976; Rich, 1977). Rutgers tomato is commonly used as an indicator plant (Figs. 4.12 and 4.13).

Tuber symptoms are more obvious and easier to detect than are plant symptoms. The disease derived its name from the typically elongated, spindly shape of affected tubers which may be more or less cylindrical or tapered at one or both ends (Fig. 4.14). Eyes are more numerous and more conspicuous than those of healthy tubers, and have pronounced "eyebrows." The skin of red varieties is tender and a lighter shade of red than the skin of healthy tubers (Folsom *et al.,* 1955; Goss, 1930; O'Brien and Rich, 1976; Rich, 1977).

Fig. 4.11.   Potato plant infected with potato spindle tuber viroid. Note upright growth, sharp angle of branching, and slight twisting.

Potato spindle tuber viroid affects not only shape but also reduces the number and size of tubers. Reduction in yield is probably influenced by the strain of PSTV and the potato cultivar involved. LeClerg *et al.,* (1944) recorded a 43.2% reduction in yield. Hunter and Rich (1964) found that yield of Saco was reduced by 64.8% when infected with PSTV (Figs. 4.15 and 4.16). Singh *et al.* (1971) reported that a mild strain of PSTV reduced yield of Saco 17–24%, while a severe strain caused a yield reduction of 64%. A tuber abnormality resembling spindle tuber was observed on the Redskin cultivar in Great Britian. At first it was suspected that the causal agent was PSTV, but Cammack (1964) concluded that this was not the case.

Fig. 4.12. Rutgers tomato plant infected with potato spindle tuber viroid.

Fig. 4.13. Close-up of rutgers tomato leaf infected with potato spindle tuber viroid (right). Leaf on left is healthy.

Fig. 4.14.   Green Mountain potato tubers infected with potato spindle tuber viroid.

B. Causal Agent

For many years the causal agent of potato spindle tuber was considered to be a virus (Folsom *et al.,* 1955; Goss, 1930). Recently, however, Diener and others have demonstrated that it is a single-stranded ribonucleic acid (RNA) which lacks a protein coat (Diener, 1975; Sogo *et al.,* 1973). Diener applied the name "viroid" to this tiniest known agent of plant disease. PSTV has a molecular weight of only 80,000–90,000 (Diener, 1975). McClements and Kaesberg (1977) claimed that electron miscroscopy indicated that PSTV is composed of two single-stranded RNA components of different sizes, one linear and one circular. They estimated the molecular weight of the linear molecule as 110,000 and the circular molecule as 137,000.

Apparently there are several strains of PSTV affecting host response and tuber yield (Singh, 1970; Singh *et al.,* 1971). For many years unmottled curly dwarf was considered to be a distinct disease, but research workers showed that it was caused by a strain of PSTV (Folsom *et al.,* 1955; Goss, 1931; Raymer and O'Brien, 1962; Rich, 1977; Smith, 1972). MacLachlan (1960) concluded that a virus of the aster yellows type is associated with

Fig. 4.15. Hill of five Saco potato tubers from plant infected with potato spindle tuber viroid (upper). Hill of eight healthy Saco potato tubers in lower portion of photo.

spindle tuber in eastern Canada. However, other workers have not confirmed his observations.

C. Host Range

Smith (1972) lists tomato *(Lycospersicon esculentum)* and *Nicotiana glutinosa* as hosts. *Nicotiana glutinosa* shows flower breaking as a symptom and adds *N. debneyi, N. rustica, N. physaloides,* and *Physalis floridana* as symptomless hosts. Species of *Capsicum, Datura,* and *Nicandra* are also symptomless hosts (O'Brien and Raymer, 1964). Eggplant *(Solanum melongena* cv. Black Beauty) is a host (O'Brien, 1972).

Hunter (1965) and Raymer (1975) used the Rutgers cultivar of tomato as an indicator host. Inoculated plants develop a bunchy top which becomes yellow, curled, or epinastic and somewhat necrotic. High temperature (30°–38°C), high light intensity, and fertile soil are essential for good symptom expression. New growth is chlorotic to white under continuous light (Yang and Hooker, 1977). Singh and Bagnall (1968) suggested the use of *Solanum*

Fig. 4.16. Left row, Saco plants infected with potato spindle tuber viroid. Right row, healthy Saco potato plants.

rostratum as a test plant for PSTV. More recently Singh (1973) has recommended the use of *Scopolia sinensis* as a good local lesion indicator. Local lesions develop under low light intensity (400 fc) and low temperature (73°). A temperature of 76°F is too high for good symptom development.

D. Transmission

Potato spindle tuber viroid is easily mechanically transmitted. It is spread by the cutting knife, by rubbing together the cut surfaces of infected and healthy seed pieces, by contaminated planting and cultivating equipment, and by leaf contact (O'Brien and Rich, 1976; Smith, 1972). Manzer and Merriam (1961) demonstrated field transmission by cultivating equipment.

Two species of aphids, *Myzus persicae* and *Macrosiphum euphorbiae,* are reported to be vectors. It is probable that other aphids can also transmit PSTV in a stylet-borne manner. Other insects recorded as vectors include grasshoppers *(Melanoplus* spp.), flea beetles *(Epitrix cucumeris* and *Systena taeniata),* tarnished plant bugs *(Lygus pratensis),* larvae of the Colorado potato beetle *(Leptinotarsa decemlineata),* and the leaf beetle *(Disonycha triangularis)* (Goss, 1931; Smith, 1972). These chewing insects are probably not very efficient vectors (Schumann *et al.,* 1980).

Potato spindle tuber viroid is transmitted through the seeds and pollen of infected potato plants (Fernow *et al.,* 1970). It is tuber-perpetuated from one season to another.

E. Control

Spindle tuber is one of the most difficult diseases to control. Growers should plant high-quality seed potatoes, free from the disease. Seed potatoes should be graded before they sprout, and the cut surfaces should be dried with a minimum of contact. Planting small, whole tubers will eliminate spread by the cutting knife.

Diseased plants should be rogued early in the season. Use of tuber-unit seed plots is an important aid in the identification of diseased plants and units. Field equipment should be disinfested. Cultivation and hilling should be completed while the plants are still small. The use of herbicides may reduce the amount of cultivation required. Insect vectors should be controlled. Aerial application of sprays or dusts should reduce field spread by tractors and sprayers (Manzer and Merriam, 1961; O'Brien and Rich, 1976; Rich, 1977; Schumann *et al.,* 1977).

Highly susceptible cultivars, such as Kennebec and Saco, should be avoided. Commercial cultivars which are resistant to PSTV are not yet available, but breeders are making progress in identifying sources of resistance (Manzer *et al.,* 1964b).

XVII. ASTER YELLOWS MYCOPLASMA

Aster yellows is a serious disease of many plants, including potatoes. The potato disease incited by the aster yellows mycoplasma (AYM) has many names or synonyms, the most common of which is purple-top wilt. Other synonyms for this or similar diseases of potato include apical leafroll, purple dwarf, purple top, purple-top roll, bunch top, blue stem, late-breaking virus, moron disease, hay wire, and possibly yellow top (Bonde and Schultz, 1953; Leach and Bishop, 1946; Martyn, 1968; Menzies, 1950b; Nagaich and Girin, 1973; O'Brien and Rich, 1976). According to Conners (1967), purple-top wilt is a symptom of primary infection, and the hay wire stage is due to secondary infection. The French refer to the disease as *touffe pourpre* (Conners, 1971).

A. Symptoms

The early symptoms of purple-top wilt develop at the apex of the plant. Young leaflets are pinched or curled at their base and developed a marked reddish or purple pigment in those cultivars which contain anthocyanin. Cultivars which lack anthocyanin (white-blossomed cultivars) become chlorotic instead of purple at the base of the leaflets.

Infected plants lose their apical dominance. Axillary shoots develop in the leaf axils, and frequently aerial tubers develop at these sites. Infected plants may wilt, become necrotic at their base and in the roots, and die prematurely.

Tubers from infected plants are often soft and spongy. They may also develop a stem-end necrosis. Infected tubers sometimes fail to sprout or they may develop spindling sprouts.

Advanced stages of purple top wilt may resemble psyllid yellows and haywire (Connors, 1967; Hodgson *et al.*, 1974; O'Brien and Rich, 1976; Rich, 1977).

B. Host Range

The aster yellows mycoplasma has a very wide host range affecting numerous species of cultivated and wild plants, including weeds. It is a serious disease of china aster and also attacks many other ornamental plants, such as daisies, delphinium, marigold, petunia, and phlox. Disease symptoms are also produced on many vegetables, including carrots, celery, lettuce, endive, parsley, and New Zealand spinach. It causes green petal of strawberry and phyllody of clover. The complete range includes more than 170 species representing 38 families of dicotyledons (Westcott, 1971).

There are numerous strains of AYM, identified principally by host reaction. Kunkel identified the eastern strain which supposedly did not infect celery and the western strain which did. At least five strains have been recognized in the United States and Canada (Walker, 1969).

C. Transmission

The causal agent is transmitted by the aster leafhopper, *Macrosteles fascifrons* (Stäl), but other species of leafhopper may also be involved.

It is not sap-transmitted and is rarely, if ever, tuber-perpetuated. Most infected tubers produce very weak, spindly sprouts or no sprouts at all.

D. Control

The most important control measure for purple-top wilt and related diseases is the effective control of the aster leafhopper vectors. Another recommended practice is to control weed hosts around the edges of potato fields which may harbor the aster yellows mycoplasma. Isolation of potato fields away from other hosts, such as clover, may also be helpful. Roguing of diseased potato plants may also be beneficial (Hodgson *et al.,* 1974; O'Brien and Rich, 1976; Rich, 1977).

XVIII. POTATO WITCHES' BROOM

Potato witches' broom or witch's broom has a worldwide distribution, but it occurs primarily in the northwestern United States and western Canada (Hodgson *et al.,* 1974; O'Brien and Rich, 1976; Wright, 1950). The French name for this disease is *balai de sorcière de la pomme de terre.* Spanish-speaking people know it as *escoba de bruja de la papa.* It is referred to as *Hexenbesenkrankheit der Kartoffel* in German (Miller and Pollard, 1977).

For many years the causal agent of this disease was thought to be a virus, but recent evidence indicates that it is a mycoplasma (Brcak *et al.,* 1969; Doi *et al.,* 1967; Harrison and Roberts, 1969; Maramorosch *et al.,* 1970). At least three strains of the organism are recognized in Canada (Hodgson *et al.,* 1974; Wright, 1952, 1954).

Holmes (1948) and Smith (1957) consider "wilding" and "semiwilding" to be synonyms of witches' broom, but Todd (1954) considers them to be separate and distinct diseases. "Northern stolbur" may also be a synonym.

A. Symptoms

Affected plants are dwarfed and chlorotic. Leaf margins are often reddish yellow. Diseased plants produce many slender, spindly stems and shoots. Leaves are simple, small and velvety in texture. Flowers and seed balls may occur in abundance. Aerial tubers may develop, which also may produce new growth.

Many tiny tubers are produced by a diseased plant. Frequently the little tubers form in chains or beads along the stolon. The small tubers often sprout the same season without a dormant period and produce an excessively large number of spindling sprouts.

B. Host Range

In addition to potato, tomato *(Lycopersicon esculentum),* alfalfa *(Medicago sativa),* and clover *(Trifolium* sp.) have been reported as hosts of the witches' broom mycoplasma (O'Brien and Rich, 1976). Probably other plants serve as hosts also even though they may not produce easily recognizable symptoms.

C. Transmission

The potato witches' broom mycoplasma is tuber-perpetuated. It is also transmissible by dodder and by grafting. It is not sap-transmissible.

Several species of leafhoppers have been reported as vectors of the causal agent of this disease. They include *Peragallia sinuata* in Europe and *Scleroracus flavopictus* in Japan (Ishihara, 1969). In western Canada *S. dasidus* and *S. balli* are capable of transmitting the causal agent from clover and alfalfa to potato but not from potato to potato (O'Brien and Rich, 1976).

D. Control

Disease-free seed potatoes and carefully roguing are recommended practices. Control of leafhopper vectors should also be practiced.

XIX. MISCELLANEOUS VIRUSES

A. Potato Stunt Virus

Potato plants infected with potato stunt virus are dwarfed, erect, and exhibit some leaf distortion and necrosis. Smith (1977) lists potato dwarf virus as a synonym. He states that gray-black, water-soaked lesions develop on the lower leaves which later collapse, wilt; and fall. The virus is tuber-perpetuated and sap-transmissible to *Nicotiana tabacum* and potato. It causes necrotic spots and rings on *N. tabacum*. The cryptogram for this virus is */*:*/*:*/*:S/* (Harrison, 1971).

B. Cucumber Mosaic Virus

This virus has a wide host range including numerous crop plants and weeds. It causes yellowing and necrosis of potato shoots. It is sap-transmissible to *Nicotiana tabacum*. The green peach aphid, *Myzus persicae,* is its most important insect vector. It is rarely tuber-perpetuated. Its cryptogram is R/1:1/18:S/S:S/Ap. It has isometric particles 30 nm in diameter (Harrison, 1971). Control measures include proper weed control and a good aphid control program.

C. Beet Curly Top Virus

This virus has a wide host range, including weeds and crop plants. Names applied to potato plants affected with this virus include potato curly top, green dwarf, and haywire (Menzies and Giddings, 1953; Milbrath, 1946; Rich, 1977; Smith, 1972). Infected potato plants are stunted with yellowish, rolled leaflets and twisted petioles. Axillary shoots may develop, and severely affected plants finally turn yellow and die. The virus is sap-transmissible with great difficulty. Its natural vector is the beet leafhopper. Control measures include proper control of weeds and insect vectors.

Diseases Caused by Nematodes and Insects

I. INTRODUCTION

Nematodes or eelworms are tiny almost invisible worms which often inhabit the soil and frequently attack the roots and other underground plant parts, including potato tubers. Parasitic nematodes can be identified by the presence of a stylet or spear. Some species attack roots or tubers directly and are primary causal agents of disease. Other act synergistically with fungi, such as *Verticillium* (Hide and Corbett, 1973; Morsink and Rich, 1968) to increase the severity of the fungus disease. Others serve as virus vectors.

Many insects attack potato plants or tubers or both, but few are directly responsible for potato diseases. However, some produce toxins which cause toxicogenic diseases. Included in this category are the potato or tomato psyllid [*Paratrioza cockerelli* (Sulc)] and the potato leafhopper [*Empoasca fabae* (Harris)] which cause psyllid yellows and hopperburn, respectively. These two toxicogenic diseases will be discussed briefly.

II. GOLDEN NEMATODE DISEASE

Golden nematodes or potato cyst nematodes have been a serious pest of potatoes in Europe for many years (Evans and Brodie, 1980). They were introduced into the United States and became a serious problem on Long Island, New York, in the mid-twentieth century. However, they have been confined almost entirely to that area through quarantines and strict sanitation practices (Sands, 1979). They are thought to be endemic in the Andes Mountains of Peru and have been reported from Vancouver Island, British Columbia, and Newfoundland.

English names applied to this disorder caused by golden nematodes are golden nematode, golden nematode disease of potato, and potato sickness. French names include *maladie de la pomme de terre* and *nématode sur pomme de terre*. Spanish names for the disease are *nemátodo dorado de la papa* and *nemátodo quiste*. It is refered to as *gelbes Kartoffelnematode* or *Kartoffelzystenälchen* in German (Miller and Pollard, 1977).

A. Causal Agents

The golden nematode, formerly classified as *Heterodera rostochiensis* Wollenweber, is usually classified as *Globodera rostochiensis* (Wollenweber) Mulvey & Stone today (Miller and Pollard, 1977; Mulvey and Stone, 1976). *Globodera pallida* also occurs on potatoes in South America and Europe (Evans and Stone, 1980; Hooker, 1981; Stone, 1972).

Races or biotypes of *G. rostochiensis* have been recognized by nematologists and plant breeders. The wild Argentine potato, *Solanum vernei* Bitt. & Wittm. ex Engl. and certain clones of the cultivated Andean potato, *S. tuberosum* L. ssp. *andigena* (Juz. & Buk.) Hawkes possess resistant genes (O'Brien and Rich, 1976).

B. Symptomatology

Injured plants are stunted, with weak and spindly stems. Leaflets are chlorotic or necrotic. Plants may wilt during the heat of the day and partially recover at night. They may develop a tufted or bunchy-top appearance. Severely affected plants never emerge or die shortly afterward, resulting in a poor stand with bare spots in a field (Fig. 5.1).

The root system is much reduced. Roots are small, fibrous, and highly branched, giving them a feathery appearance. Tubers are reduced both in size and number. Yield loss is influenced by soil type, moisture, and level of infestation. Dry weather and light soil favor heavy nematode infestation.

The signs include the presence of *G. rostochiensis* eelworms and the golden cysts which give the nematode its name (Fig. 5.2). The golden brown to dark brown cysts contain many eggs which later hatch into nematode larvae and attack the host plants. The causal agent is transported from place to place in the form of small egg-containing cysts which may be present in contaminated soil, containers, equipment, or clinging to the surface of potato tubers.

Fig. 5.1. Potato field showing injury from golden nematodes. Note poor stand and weakened plants. (Photo courtesy of W. F. Mai, Cornell University.)

C. Control

Strict sanitation practices and the enforcement of plant quarantines are absolutely essential for containment of this disease. Other important control measures include soil surveys to determine the presence or absence of golden nematodes, washing potatoes and marketing them in new paper bags, steam sterilization of contaminated equipment, and long crop rotations without potatoes and tomatoes for about 8 years. Soil and nursery stock from contaminated areas should not be moved to uncontaminated areas. Infested soil should be fumigated with D-D, EDB, Vorlex, or some other effective nematicide (Wilson, 1968). Resistant cultivars such as Hudson, Peconic, or Wauseon should be selected for planting in areas where soil is infested (Chitwood, 1951; Cunningham *et al.,* 1968; Mai and Lear, 1953; Mai and Spears, 1954; O'Brien and Rich, 1976; Peterson and Plaisted, 1966; Plaisted *et al.,* 1973; Rich, 1977; Sand, 1979). Newer cultivars and seedlings which are resistant to golden nematode injury are Atlantic, Belchip, Campbell 11, Campbell 13, Rosa, AF186-5, AF201-25, AF205-9, B8943-4, C7358-26A, CA02-7, and CF7523-1 (Murphy *et al.,* 1982).

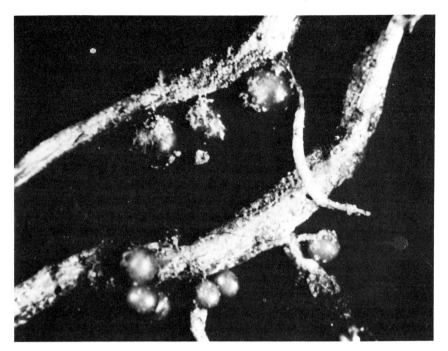

Fig. 5.2.   Golden nematode cysts on potato roots. (Photo courtesy of W. F. Mai, Cornell University.)

III. LESION NEMATODE ON POTATO

Reports of damage to potatoes by several species of lesion nematodes *(Pratylenchus* spp.) occur in the literature. In some cases the damage is direct (Miller and Pollard, 1977), while in others it is indirect or synergistic (Rich, 1977).

English names include lesion nematode or meadow nematode on potato. French names are *nématode des lésions de racines* and *des prairies sur pomme de terre.* It is called *nemátodo de las lesions radiculares* (Miller and Pollard, 1977) in Spanish.

A. Causal Agents

According to Miller and Pollard (1977) *Pratylenchus scribneri* can cause direct damage to potato tubers. Other species of *Pratylencus,* probably *P. steineri,* can also cause direct damage to potato tubers.

Pratylenchus penetrans (Cobb) Chitwood & Oteifa has a wide host range, including potatoes (Dickerson *et al.,* 1964; Mai *et al.,* 1977; Morsink, 1966).

These lesion or meadow nematodes are migratory and endoparasitic, approximately 0.4–0.7 mm long and 20–25 μm in diameter. The body is stout and cylindroid, the head is blunt and strong with a stout spear, and the tail is bluntly rounded (Agrios, 1978).

A synergistic action between *P. penetrans* and *Verticillium alboatrum* was demonstrated by Morsink (1966), and Mai *et al.* (1977) reported that onset of disease symptoms was hastened by this interaction. Several other workers have shown that soil fumigation decreases nematode populations and/or severity of Verticillium wilt and increases potato yields (Hawkins and Miller, 1971a,b; Kunkel and Weller, 1966; Miller and Edgington, 1962; Miller and Hawkins, 1969). It is highly probable that other fungi invade roots through the wounds or lesions made by this nematode.

B. Symptomatology

Miller and Pollard (1977) report that small pimplelike bumps scattered over the tuber surface can be caused by *Pratylenchus scribneri*. They also claim that other species of *Pratylenchus* cause small light to dark brown-purple pimples on potato tubers. Affected tubers gradually lose their firmness, shrink, and may become dry and mummified.

Herbaceous plants which are affected by lesion nematodes usually appear stunted and chlorotic. The injury may be mistaken for nutrient deficiency or drought. Root symptoms are more definitive consisting of distinct lesions on the roots and root hairs. Lesions are elongate, water-soaked, and light colored at first but soon turning dark brown to almost black. They may coalesce and may girdle the roots. The root system is dark, stubby, and much reduced. Secondary organisms, especially fungi, often invade the root lesions (Agrios, 1978).

Signs consist of the nematode eggs, larvae, and adult nematodes (Fig. 5.3), and are both ecto- and endoparasitic.

C. Control

Soil fumigation appears to be the most effective and practical method of controlling lesion nematodes. Although it does not completely eradicate them, it reduces their population to the point that injury is considerably reduced. Fall treatment with D-D, EDB, Vorlex, or some other effective fumigant is recommended. Twenty gallons of D-D or 10 gal of Vorlex per acre are suggested by Miller and Edgington (1962). A combination of 5 gal of Picfume (chloropicrin) and 20 gal of Telone per acre proved effective in trials by Kunkel and Weller (1966). Hawkins and Miller (1971a,b) concluded that row treatment with Vorlex or aldicarb was effective for only 1

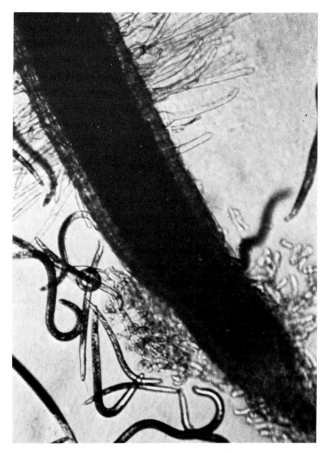

Fig. 5.3.   Lesion nematodes attacking strawberry root. The injury to potato roots is similar.

year, while fall fumigation of the entire area was effective for 3 years (Miller and Hawkins, 1969).

Summer fallow is helpful in hot, dry climates. Crop rotation is rather ineffective due to the wide host range of these nematodes (Agrios, 1978).

IV. POTATO-ROT NEMATODE DISEASE

The potato-rot nematode has been known for many years in Europe and is a serious pest in some regions. The first report of the occurrence of this pest parasitizing potatoes in North America was from Prince Edward Island, Canada, where it was limited to only a few farms. Since that time it has been reported from British Columbia, Canada, and Idaho and Wisconsin in the United States, but it is not widespread.

The English literature contains the names potato-rot nematode and potato-rot nematode disease for the nematode and the disease which it causes. The French call it *nématode de la pourriture du tubercule de la pomme de terre.* Spanish-speaking people call it *nemátodo de la pudrición de la papa, pudrición seca de la papa, nemátodo del tubérculo,* or *nemátodo de la podredumbre de la papa* (Miller and Pollard, 1977).

A. Causal Agent

Potato-rot nematode disease is caused by the potato-rot nematode *Ditylenchus destructor* Thorne (Thorne, 1945). Some of the older literature may list *D. dipsaci* as the scientific name of this nematode.

B. Symptomatology

Above-ground symptoms have not been associated with this disease in the United States. However, European workers have reported that leaves tend to be small with a yellowish tinge and that stems are thickened and cracked.

Early tuber symptoms consist of small holes resembling wire worm injury. These enlarge to gray patches which become dry, granular, and cracked. A grayish or brownish granular decay develops in the tubers. It may be visible at harvest time or may develop during storage and can be easily mistaken for Fusarium dry rot while the cracking resembles symptoms of bacterial ring rot.

Signs consist primarily of the numerous, long eel-like nematodes with stylets. Other stages include the eggs and larvae of these nematodes. The nematodes overwinter as adults or larvae. They feed mainly on fungi and can survive in fallow soil for several years. Saprophytic nematodes may also be present, but they lack stylets (Hodgson *et al.,* 1974; O'Brien and Rich, 1976; Rich, 1977).

C. Control

Only disease-free, nematode-free, certified seed potatoes should be planted. Infested fields should be fumigated with a good soil fumigant such as D-D, EDB, or Vorlex (Blodgett, 1943; Blodgett and Rich, 1950; Darling, 1957; O'Brien and Rich, 1976; Rich, 1977; Thorne, 1945). Hodgson *et al.*

(1974) suggest crop rotation with grains, grasses, or alfalfa, but not with clover.

V. ROOT KNOT NEMATODE DISEASE

Root knot nematodes are known to attack a great variety of crop plants and weeds. They can be a serious problem, especially in warm soil climates such as the southern and western states of the United States and in greenhouses.

English-speaking people refer to the disease simply as root knot or root knot nematode disease. French names for the disease are *nodosité des racines* or *nématode cédidogène sur pomme de terre*. Spanish names include *nemátodo nodular, nemátodo del nudo de la raíz, mezquino de la papa,* and *nemátodo de los nudos radiculares*. The German word for it is *Wurzelgallenälchen* (Miller and Pollard, 1977).

A. Causal Agent

Root knot is caused by *Meloidogyne* spp. Miller and Pollard (1977) cite *M. acroneta* Coetzee as the causal agent. Others (Mai *et al.,* 1960; Sasser, 1954) list *M. hapla* Chitwood and *M. incognita* (Kofoid and White) Chitwood as the causal agents. Another name for this pathogen, often found in the older literature, is *Heterodera marioni* (Cornu) Goodey (Cunningham 1936; Cunningham and Mai, 1947).

Adult males of *Meloidogyne* are wormlike, about 1.2–1.5 mm long and 30–36 μm in diameter. Adult females are pear-shaped, about 0.40–1.30 mm long by 0.27–0.75 mm wide (Agrios, 1978).

B. Symptomatology

Early symptoms occur as small galls on the roots, followed by excessive root branching. Infected plants become stunted and tend to wilt in hot, dry weather. Small to large pimples or warts develop on the surface of infected tubers (Fig. 5.4). Small brown "fly specks" often occur in affected tissue, usually between the skin and the vascular ring. Yields are reduced (Miller and Pollard, 1977; O'Brien and Rich, 1976; Rich, 1977).

Signs consist of eggs, larvae, eelworms with stylets, and pear-shaped mature females. The eggs overwinter in the galls on the roots (Hodgson *et al.,* 1974; O'Brien and Rich, 1976).

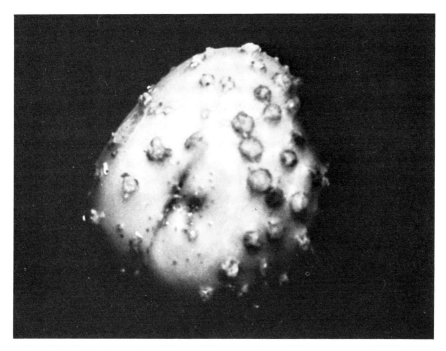

Fig. 5.4. Potato tuber injured by root knot nematodes. (Photo courtesy of W. F. Mai, Cornell University.)

C. Control

Control measures include crop rotation with cereals and grasses and soil fumigation (Fig. 5.5) with an effective nematicide such as D-D, DBCP, EDB, or Vorlex (Wilson, 1968; Winstead *et al.* 1958). Aldicarb, oxamyl, and phenamiphos also show considerable promise (Agrios, 1978; Rodervald, *et al.* 1975). Other recommended practices such as use of disease-free seed potatoes and weed control should be practiced (Hodgson *et al.,* 1974; O'Brien and Rich, 1976).

VI. RENIFORM NEMATODE DISEASE

The reniform nematode has been reported to parasitize potatoes in Egypt, India, and the United States (Abdel-Rahman *et al.,* 1974; Rebois *et al.,* 1978; Rebois and Webb, 1979).

Fig. 5.5.   Tractor equipped with soil fumigation attachment for control of nematodes.

A. Causal Agent

Reniform nematode disease is caused by *Rotylenchus reniformis*. This nematode has a wide host range.

B. Symptomatology

Aboveground plant parts usually remain symptomless. Slight dwarfing, lack of vigor, and/or early maturity might be expected where nematode populations are extremely heavy. Affected tubers are often misshapen with excessive corkiness or suberization of the skin, accompanied by a reduction in yield (Rebois *et al.*, 1978).

Signs consist of the presence of nematodes and egg masses on the roots of affected plants. They are rarely found on tubers even though the nematodes feed on the tubers (Rebois *et al.*, 1978).

C. Control

Abdel-Rahman *et al.* (1974) reported that aldicarb decreased nematode populations and increased yield and quality of potatoes. The La Rouge cultivar is resistant to *R. reniformis* (Rebois and Webb, 1979).

VII. HOPPERBURN

Hopperburn is a serious toxicogenic disease of potato. It is caused by feeding of the potato leafhopper, *Empoasca fabae* (Harris). This insect is small, light green, and fragile-looking. Adults are about 1/8 inch long, wedge-shaped, with long hindlegs. They can hop or fly from plant to plant and migrate long distances. They do not overwinter in the northern cold climates but migrate to northern potato fields from where they overwinter in the southern states (Schulz, 1976; Simpson, 1977).

A. Symptomatology

Symptoms of hopperburn are browning of the tips and margins of potato leaflets following the feeding of the toxicogenic nymphs or adults of *E. fabae*. Only a narrow strip along the midrib of severely affected leaflets remains green. Petioles are shortened which causes a crowding of the leaflets (O'Brien and Rich, 1976; Schulz, 1976).

B. Control

Hopperburn can be adequately controlled by controlling the toxicogenic potato leafhoppers which cause it. A preplant soil application of a systemic insecticide, such as disulfoton, should give early season control. If necessary, this can be followed by mid- and late-season application of foliar insecticides (O'Brien and Rich, 1976; Schulz, 1976; Simpson, 1977).

VIII. PSYLLID YELLOWS

Psyllid yellows is caused by feeding of nymphs of the toxicogenic potato psyllid, *Paratrioza cockerelli* (Sulc.). The adult insects overwinter on plants belonging to the nightshade family in Texas and New Mexico. They are about 1/10 inch long and hold their wings over their body like a tent when resting. Eggs are pale yellow to orange, spindle-shaped, and suspended on short stalks. Nymphs are light yellow or green, scalelike, and remain rather inactive on the under sides of the leaves where they feed and secrete their toxin. Most damage is caused to potatoes in Colorado, Nebraska, and New Mexico (Simpson, 1977).

A. Symptomatology

The first symptoms of this toxicogenic disease are yellowing of the margins and an upward rolling of the basal portions of the small leaflets on young leaves of injured plants. The rolled leaves turn reddish, purplish, or yellow, depending on the pigments in the specific cultivar. Older leaves of severely affected plants roll upward, making the entire plant appear reddish, purplish, or yellowish. Certain stages of the disease resemble symptoms caused by *Rhizoctonia*. Other stages resemble symptoms of leafroll, purple top or other viral diseases. Brown necrotic areas develop on the leaves resulting in their eventual death. Axillary buds develop and give the plant a compact, pyramidal shape. Aerial tubers may develop. They often produce leafy shoots or rosettes of small, yellow leaves.

Plants affected with psyllid yellows produce an abnormally large number of small tubers many of which are too small to be of any value. They may sprout and produce new plants before maturing. Unlike viral diseases, psyllid yellows is not tuber-perpetuated from one year to the next (O'Brien and Rich, 1976).

B. Control

Psyllid yellows can be rather easily controlled by controlling the potato psyllid which causes it. Systemic insecticides, such as aldicarb, disulfoton, or phorate, applied in the furrow at planting time will give adequate control during the early growing season. If necessary foliar applications of demeton, diazinon, dimethoate, endosulfan, methomyl, oxydemetonmethyl, or parathion are suggested (Simpson, 1977). If one adult is collected with 100 sweeps of a sweep net, it is time to make the first foliar application.

Noninfectious Diseases

I. INTRODUCTION

In addition to infectious diseases caused by bacteria, fungi, viruses, viroids, mycoplasmas, and nematodes, there is a large group of noninfectious diseases or disorders caused by a variety of environmental factors. Sometimes they are referred to as abiotic or nonparasitic diseases. The "Index of Plant Diseases in the United States" (Anonymous, 1960) lists 40 disorders which are either noninfectious or due to unknown causes. Some new disorders, such as air pollution damage, have been recognized since that time. Smith and Wilson (1978) describe about twenty nonparasitic disorders and injuries to potato tubers.

II. AIR POLLUTION INJURY

A. Symptoms

Mild symptoms consist of a chlorotic or necrotic spotting of the upper leaf surface. The name "speckle leaf" has been used in Michigan for this disorder (Hooker *et al.,* 1972; 1973). Severe symptoms include more pronounced chlorosis, bronzing, and necrosis of leaves resulting in early maturity of affected plants and a significant reduction in yield. Dead leaves remain attached to the stem. Leaf necrosis occurs first in the palisade mesophyll cells and advances to the spongy mesophyll cells beneath (Hooker *et al.,* 1972; 1973). Under greenhouse conditions in Beltsville, Maryland,

149

a silvering of the lower leaf surface of Green Mountain potato plants was observed (Fig. 6.1) (A. E. Rich, unpublished data). Air pollution injury appears to favor leaf-spotting fungi such as *Alternaria, Chaetomium,* and *Stemphylium* which thrive on senescent leaves (O'Brien and Rich, 1976).

B. Cause(s)

As the name implies, air pollution injury is caused by one or more air pollutants. Ozone appears to be the worst offender (Brennan *et al.,* 1964; Hooker *et al.,* 1972; Rich and Hawkins, 1970). Probably other air pollutants, such as sulfur dioxide and PAN (peroxyacyl nitrate), also contribute to the problem in some areas.

C. Control

Potato growers should avoid areas where air is heavily polluted with ozone or other pollutants. If in a marginally polluted area they should plant tolerant cultivars and avoid cultivars known to be highly sensitive to air pollutants. Reports on tolerance and sensitivity vary from one geographic area

Fig. 6.1.   Air pollution injury to Green Mountain potato leaves. Ozone was the principal pollutant.

to another, possibly due to a difference in the types or concentrations of pollutants present. Under Maryland conditions, Haig, Irish Cobbler, and Norland were found to be sensitive; Avon, Peconic, and Superior were intermediate; and Alamo and Kennebec were tolerant (Heggestad, 1970). In Michigan, Haig, Norchip, Norgold Russet, and Superior were most sensitive, while Katahdin, Kennebec, and Sebago were most tolerant (Hooker *et al.,* 1972). Studies in Connecticut by Rich and Hawkins (1970) showed that Alamo, Alaska, Bake King, Katahdin, Kennebec, Lenape, and Wausean were more sensitive than Houma, Ona, Peconic, and Superior.

III. BLACKHEART

A. Symptoms

There are no vine symptoms of blackheart, and external tuber symptoms are rare. They consist of moist purplish, brown, or black areas or patches on the surface of affected tubers. The most prevalent symptom, from which this disease derives its name, is a large gray, purplish, or black discolored area in the center of the potato tuber (Fig. 6.2). It is usually restricted to

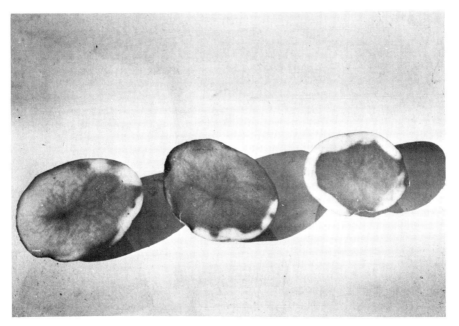

Fig. 6.2. Potato tubers cut to show blackheart symptoms.

the heart of the potato but it may radiate to the skin. The discolored tissue is usually sharply delineated from the surrounding healthy tissue. The affected tissue may dehydrate, shrink, and cause cavities in advanced cases or under long storage conditions (Blodgett and Rich, 1950; O'Brien and Rich, 1976).

B. Cause

Blackheart is due primarily to asphyxiation. This is usually a result of high-temperature storage and lack of ventilation. It used to occur rather frequently in freight cars, especially when they were heated to prevent the potatoes from chilling (Blodgett and Rich, 1950; O'Brien and Rich, 1976).

According to Smith and Wilson (1978) it can also occur in the field when the soil is flooded or the soil temperature is extremely high. They also claim that prolonged storage near 0°C will induce blackheart.

C. Control

Blackheart can usually be controlled by proper ventilation and temperature control of storage sheds and freight cars. The temperature in heated cars should not be allowed to rise above 16° to 21°C. Potatoes should not be stored in deep piles without forced ventilation through the piles. Tubers should be removed from hot, dry soils as soon as the vines die and should not be allowed to remain on hot, dry soils after digging (O'Brien and Rich, 1976).

There appears to be some difference in susceptibility between cultivars, but none are known to be immune. Irish Cobbler and Russet Burbank are somewhat resistant, while Chippewa is much more susceptible (A. E. Rich, unpublished data).

IV. BRUISES

Bruising of potato tubers is a serious problem. It results in grade defects, poor appearance, poor keeping quality, and a reduction in marketability. Bruises also provide an avenue of entrance for bacteria and fungi, especially *Fusarium* spp., which cause the tubers to rot.

A. Symptoms

Symptoms consist of broken skin and crushed or shattered tissue (Fig. 6.3) where the injured tubers came in contact with some hard object. The injured tissue turns pink at first and then turns gray or brown. (Internal black spot will be treated separately.)

Fig. 6.3. Bruised and cracked potato tubers due to rough handling. (Photo courtesy of Maine Life Sciences and Agricultural Experiment Station.)

B. Cause(s)

Bruising is caused primarily by rough handling during the harvesting, storing, grading, shipping, or other handling process. Sometimes delivery men and produce clerks are responsible for handling potatoes too roughly after they have been carefully graded and packed.

C. Control

Careful handling should start with the harvesting process. Potatoes should not be harvested when they are cold. Digging should be delayed on cold mornings until the soil has warmed up. Potato harvesters should be adjusted properly and run at the appropriate speed so that potatoes roll gently but do not bounce or drop unnecessarily. Sufficient soil should be carried on the potato digger bed to cushion the tubers.

Potatoes should drop gently from the harvester into the truck, tractor trailer, or other container. Then they should be dumped or conveyed gently into storage by means of properly adjusted chutes or bin loaders. All equipment used in harvesting, storing, and grading should be padded properly with rubber, neoprene, or other material to reduce tuber bruises. The Maine Extension Service has conducted a rigorous antibruise campaign to minimize tuber bruising (Stiles, 1979).

Packaged potatoes should be handled very gently in transit, storage, and in the market and should never be thrown or dropped.

Some cultivars bruise more easily than others, but all cultivars are quite susceptible. Selecting a cultivar which is somewhat bruise-resistant may help to a certain extent, but it should not be used as a substitute for careful handling.

V. COLD TEMPERATURE INJURY

A. Symptoms

Both plants and tubers are subject to cold temperature injury. Although plants will withstand temperatures slightly below 0°C, they will suffer frost or freezing damage if the temperature drops much below 0°. Young plants will freeze back to ground level. The green tissue turns brown and gradually shrivels and dries up (Fig. 6.4). This is a problem where winter potatoes are grown, in Florida, for example. Sometimes it is a problem in the northern states and Canada when those areas have an unexpected late spring frost. Usually the plants will recover by producing new shoots, but growth is retarded.

Frosts in late summer or fall sometimes kill the vines, especially in areas with short growing seasons. This is a problem if they occur before the tubers have reached the desired size. However, they may be beneficial if the tubers have sized properly. Vine killing by frost reduces late season spread of viruses and late blight, makes harvesting easier, and eliminates the necessity of using chemical vine killers. However, if green vines are killed suddenly by frost, tubers may develop vascular discoloration resembling that caused by chemical vine killers (Rich, 1950).

Sometimes potato tubers or portions of them freeze in the field before harvesting (Fig. 6.5). Usually only the side or end of the tuber near the soil surface is affected, while the major portion of the tuber is buried with sufficient soil to prevent freezing. The tissue of the injured side or end collapses and usually dries to a flat, starchy surface area while the tissue beneath remains normal. Under severe cold weather conditions the entire crop may freeze and will have to be abandoned.

Tubers are subject to several types of low-temperature injury, the symptoms depending on how low the temperature falls, how fast it falls, and how long the tubers are exposed to the low temperatures.

Fig. 6.4. Young potato plant which has been frozen.

Fig. 6.5. Potato tuber which has been frozen.

If the temperature hovers just above or around 0°C for an extended period of time, starch is converted to sugar and potatoes usually develop a sweet taste. Usually there are no other symptoms in this temperature range. If the temperature drops slightly below 0°C the flesh of injured tubers may turn reddish brown or gray (Fig. 6.6). This has been referred to as mahogany browning (Smith and Wilson, 1978). Some cultivars are more susceptible than others.

Potatoes in storage or transit will withstand temperatures of −1° or −2°C (28°–30°F). If the temperature falls below −2°C for any length of time, serious injury will occur. Affected tubers may exhibit a net type of necrosis or a very serious necrosis and breakdown of tissue in the vascular ring area. The next stage is complete collapse or breakdown of frozen tubers. The cells rupture, collapse, and leak, and the potatoes usually become wet and messy. Bacteria often invade the frozen potatoes after they thaw and a foul odor develops (O'Brien and Rich, 1976; Richardson and Phillips, 1949; Smith and Wilson, 1978).

B. Control

As indicated earlier, plants usually recover from spring frosts even though growth is delayed. Some cultivars, such as Sebago, are more tolerant than others and appear to be adapted for winter planting in the southern states.

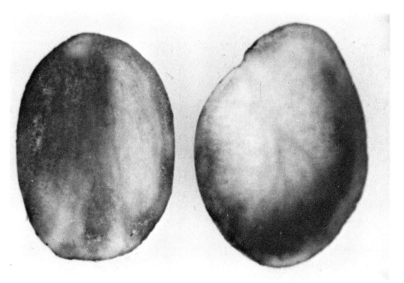

Fig. 6.6. Potato tubers cut to show symptoms of mahogany browning due to low temperatures in storage.

Dearborn (1964, 1969) developed the Alaska Russet, Alaska Frostless, and other cultivars adapted to Alaska growing conditions. Covering the Frostless young plants with soil ("hilling") is a common practice to protect them from late spring frosts. Proper hilling will also protect tubers from field frost in the fall.

Potatoes should be stored at temperatures above 2°C (36°F) to prevent the conversion of starch to sugar, resulting in a sweet taste. This will also serve as a protection from mahogany browning, tuber necrosis, or freezing and collapse. If tubers develop a sweet taste, they should be stored at about 15°C (60°F) for 1 or 2 weeks. The sugar may be utilized during respiration.

Storage bins, trucks, and freight cars should be insulated and heated in cold weather. Potatoes should never be exposed to below freezing temperatures.

VI. CRACKING

There are two distinct types of tuber cracking. One occurs before harvest and the other occurs during storage or in the market. The causes are entirely different.

A. Symptoms

Under very rapid growing conditions, tubers may crack open in the soil forming rather deep cracks or fissures. These cracks usually heal rapidly and are covered with more or less normal skin or epidermal tissue.

The other type of cracking which occurs during storage or in the market produces crescent-shaped or "thumb-nail" cracks on the tuber surface which extend for a short distance into the tuber flesh. The external surfaces of these cracks do not heal, but the exposed flesh suberizes or corks over. However, tubers lose moisture through these cracks and may become soft and flabby (Blodgett and Rich, 1950; Hodgson et al., 1973; Smith and Wilson, 1978).

B. Causes

Growth cracks are a result of rapid spurts of growth, usually caused by an irregular water supply. Fertilization may also be a contributing factor. When tubers enlarge rapidly, they may develop knobs (second growth) or they may crack open due to internal pressure.

Thumb-nail cracks are caused by dry storage conditions and rough handling of potato tubers, especially when they are cold and crisp. Combined internal pressure and bruising are significant factors contributing to this type of cracking (Blodgett and Rich, 1950; Hodgson *et al.,* 1973; Smith and Wilson, 1978).

C. Control

Growth cracks can be reduced by maintaining uniform growing conditions. Proper fertilization and irrigation are important factors. Choice of cultivars which lack the tendency to crack will also be beneficial (Murphy, *et al.,* 1982).

Thumb-nail cracks can be reduced in number and severity by warming cold tubers before they are handled, by careful handling, and by maintaining the proper humidity in storage. One should avoid cultivars which tend to crack excessively (Blodgett and Rich, 1950; Hodgson *et al.,* 1974; Murphy, *et al.,* 1982; Smith and Wilson, 1978).

VII. ENLARGED LENTICELS

A. Symptoms

Small lenticels or breathing pores are scattered over the surface of potato tubers. Under normal conditions these lenticels are small and inconspicuous. However, at times they become enlarged, corky, and resemble small kernels of popped corn or small scab lesions.

B. Cause

Enlarged lenticels are caused by excessive moisture in the soil surrounding the tubers before harvest. Storing tubers at excessively high humidity may also encourage the development or continuation of enlarged lenticels.

C. Control

Potatoes should not be overwatered during the growing season, and adequate drainage should be provided. Potato tubers should be removed from wet soil as soon as practical after completion of growth. Potatoes should be stored at moderate but not high relative humidity (Blodgett and Rich, 1950; O'Brien and Rich, 1976; Smith and Wilson, 1978).

VIII. FEATHER AND SCALD (BROWNING)

A. Symptoms

When immature tubers are harvested, the thin, fragile epidermis is torn easily, and portions of it peel away from the tubers, giving them a feathery appearance. The subepidermal surface beneath the feathered epidermis usually turns brown, giving it a scalded appearance—thus the names "scald," "browning," and "chicken skin." Severely affected tubers may lose enough moisture to make them soft and spongy. Under high humidity bacteria or fungi may attack the injured tubers, producing a sticky bacterial slime or a mold on the tubers (Blodgett and Rich, 1950; O'Brien and Rich, 1976; Smith and Wilson, 1978).

B. Cause

Feathering is caused by rough handling of immature tubers while the epidermis is soft, fragile, and scuffs easily. This can occur during harvesting, grading, shipping, or in the market. It is confined almost entirely to the early crop which is harvested while the vines are still green and the tubers are very immature.

Scald usually follows feathering. It is favored by hot, drying winds during harvesting and transportation to storage or market (Rose and Fisher, 1940).

C. Control

Harvesting immature tubers should be avoided if possible. Killing the vines several days before harvest will allow the epidermis of the tubers to mature or "set," thus making them less susceptible to feathering. Immature tubers should be handled as gently as possible to reduce scuffing of the tender skin.

Harvested tubers should be covered in transit. They should be stored in a fairly cool, shaded place under moderately humid conditions.

IX. FERTILIZER BURN

A. Symptoms

Fertilizer burn may result in seed-piece decay and missing hills. Plants which do emerge may be dark green and stunted. Roots may exhibit various degrees of necrosis.

B. Cause

Seed pieces or roots of young plants which come in contact with an excessive amount of chemical fertilizer, either dry or in solution will be likely to develop fertilizer burn.

C. Control

Fertilizer should be placed in bands on either side of the seed pieces. Excessive amounts of highly soluble fertilizers should be avoided. Applying some of the fertilizer as a side dressing may be helpful (O'Brien and Rich, 1976).

X. GREENING

A. Symptoms

Potato tubers develop a light to dark green color either before or after harvest. If the green color develops before harvest it is usually confined to one end or one side of the tuber (Fig. 6.7). The skin is dark and green and the

Fig. 6.7. Potato tuber showing greening of apical end. (Lower portion of tuber is infected with *Rhizoctonia* sclerotia.)

flesh is green or yellowish-green (Folsom, 1949). A light green color usually develops over the entire surface of tubers exposed to light after harvest. Other names applied to this disorder are sunburn or virescence.

B. Cause

The cause of greening is exposure to sunlight, indirect daylight, or artificial light (Larsen, 1949; Liljemark and Widoff, 1960). Exposure to sunlight usually takes place before harvest when tubers are not hilled properly. Exposure to indirect light or artificial light can take place during storage, in transit, on display stands, or in the home. Chlorophyll and solanine develop in the skin and flesh of tubers exposed to light for extended periods of time. Solanine ($C_{52}H_{93}NO_{18}$) is poisonous (Hansen, 1925), and imparts a bitter taste to green tubers.

C. Control

Proper hilling is the most important method of control of greening in the field. This protects the tubers from exposure to sunlight. Control of *Rhizoctonia* is also helpful.

Harvested potatoes should be stored in complete darkness. Washed potatoes turn green more rapidly than unwashed tubers. Marketing tubers in light-proof bags will protect them from turning green (Lutz *et al.,* 1951). Potatoes on display should be rotated frequently and should not be exposed to bright natural or artificial light, especially fluorescent light.

White varieties such as White Rose, Kennebec, and Katahdin are especially susceptible to greening. Breeders are attempting to develop new cultivars which will be resistant to greening (Akeley *et al.,* 1962a; Blodgett and Rich, 1950; Hodgson *et al.,* 1974; O'Brien and Rich, 1976; Smith and Wilson, 1978).

XI. HEAT AND DROUGHT NECROSIS

A. Symptoms

Tubers injured by heat and drought may exhibit a golden-yellow to brown or slate-gray discoloration in the vascular ring and other water-conducting vessels. Discoloration is often most pronounced near the stem end of affected tubers, but may occur at the bud or eye end also. There are no vine symptoms except for possibly associated wilting. As there are no external tuber symptoms, this tuber disorder is discovered only when affected tubers are cut (Blodgett and Rich, 1950; O'Brien and Rich, 1976; Smith and Wilson, 1978).

B. Cause

This problem appears to be associated with leaving tubers in hot, dry soils after the vines begin to die. As the name implies, it is associated with heat and drought and is likely to occur in irrigated sections after irrigation ceases and vines are allowed to wilt and die.

C. Control

The soil should be kept cool, moist, and shaded by green vines. The crop should be harvested as soon as the vines are dead if the soil is hot and dry. Some cultivars, such as Russet Burbank, are more tolerant of heat and drought than are others.

XII. HERBICIDE INJURY

A. Symptoms

Leaves injured by selective herbicides are usually twisted, distorted, crinkled, and/or curled, and the tips are pointed. Yields may be reduced. The color of red tubers may be affected.

B. Cause

Selective herbicides are sometimes used for weed control in potatoes. If too much chemical is used or if the plants are unusually sensitive, the injury may occur. If less selective or nonselective herbicides are used for brush control or general weed control on a windy day, they may drift to potato fields and cause severe injury.

C. Control

Herbicides should be used with great care. Careful attention must be paid to using the correct concentration and amount per acre. Application equipment must be calibrated accurately. Herbicides should not be applied on a windy day, and general broad-spectrum herbicides should not be applied on or adjacent to potato fields.

XIII. HOLLOW HEART

A. Symptoms

The term "hollow heart" is used to describe a rather conspicuous cavity which occurs at or near the center of a tuber. It is much more prevalent in large tubers than in medium-sized ones. Its occurrence in small tubers is very rare. There are no external tuber symptoms, so the disorder goes undetected until an affected tuber is cut. There may be a simple cavity surrounded by corky tissue or it may be star-shaped or contain two or more connected compartments. There are no plant symptoms.

B. Cause

Hollow heart is caused by very rapid growing conditions which are conducive to the production of oversized tubers (Dinkel, 1960; Levitt, 1942). Overfertilization, excessive rainfall or irrigation, and too wide spacing of plants can contribute to hollow heart. Small tubers can develop hollow heart due to conversion of starch to sugar under moisture stress followed by a rapid influx of water into the tubers (Crumbly *et al.,* 1973).

Kunkel (1972) and Smith and Wilson (1978) described another type of hollow heart due to excessive heat. The cavities are smaller, darker, and nearer the surface than is typical of hollow heart. Kunkel (1972) applied the name "cat's eye" to this type of injury.

C. Control

Uniform stands of closely spaced plants are important for control of hollow heart. Overfertilization and excessive irrigation should be avoided, and vines should be killed when tubers have reached their optimum size (Blodgett and Rich, 1950; Hodgson *et al.,* 1974; O'Brien and Rich, 1976; Rich, 1977; Smith and Wilson, 1978). The Cascade and Norchip cultivars are reported to be resistant (Hoyman, 1970; Johansen *et al.,* 1969a).

XIV. INTERNAL BLACK SPOT

A. Symptoms

Internal black spot is usually typified by the development of light gray, bluish, or coal-black lesions under the skin of otherwise normal potato tubers (Sawyer, 1958). According to Smith and Wilson (1978) it was first described in England in 1913. Probably it is worldwide in distribution. Other names for this disorder include bruise, blue discoloration, bluing, and blue spotting (Smith and Wilson, 1978), due to the bluish hue of the affected area.

The discolored area usually occurs from $\frac{1}{16}$ to $\frac{1}{4}$ inch below the skin surface, between the skin and the vascular ring. Occasionally it may penetrate to a depth of about $\frac{1}{2}$ inch. The bruised areas are pink at first then turn dark red 4 to 6 hr after bruising. Six to 12 hr later they become coal black, gradually fading to varying shades of gray (Smith and Wilson, 1978).

External symptoms may or may not be visible on bruised tubers. Symptoms are commonly expressed as flat or sunken more or less circular areas. The tissue underneath is starchy or corky. This phase of the disorder is generally referred to as pressure bruise or pressure spot (Smith and Wilson, 1978).

Pressure bruise and internal black spot occur most frequently on the basal half of the tuber. They are most prevalent on the shoulders of the tubers where bruises occur most frequently. There are no plant symptoms, since the causal factor does not occur prior to harvest.

B. Cause

Internal black spot is caused by bruising of tubers which is severe enough to injure the tissue beneath the skin but not severe enough to break the skin. Bruising can result from dropping tubers onto a hard surface or onto each other during harvesting, grading, packaging, shipping, or other handling procedures (Jacob, 1959).

C. Contributing Factors

There appear to be many factors which contribute to the occurrence and severity of internal black spot (Rich, 1977; Smith and Wilson, 1978). These include nutrition, date of planting, choice of cultivar, soil moisture, maturity of tubers, temperature, and carbon dioxide content of the atmosphere during storage. According to Hesen and Kroesbergen (1960) and Oswald and Lorenz (1957), high starch and low potassium content of tubers predispose them to this disorder. Other workers (Kunkel and Dow, 1961; Kunkel and Gardner, 1959, 1965; Sawyer and Collin, 1960) have shown a correlation between lack of tuber turgor and increased black spot. Howard *et al.* (1961) reported a positive correlation between increased CO_2 and black spot injury to tubers.

D. Control

The most important control measure is to avoid bruising. Potatoes should be handled only when warm, should be handled very gently, and should not be piled too deeply.

Cultivars should be planted which are not easily bruised, such as Cascade, Nampa, Oromonte, and Pontiac; Ontario and Teton are highly susceptible (Jacob, 1959; Sawyer, 1958; Twomey *et al.*, 1968).

Adequate potash should be provided (Dwelle *et al.*, 1975). Potatoes should not be allowed to lose moisture and become flaccid. They should not be allowed to dry out excessively in the field or during storage.

XV. INTERNAL BROWNING

A. Symptoms

As the name implies, normally white tuber tissue turns brown (Fig. 6.8). Several different names have been used by investigators in an attempt to differentiate between different types of internal browning, possibly due to

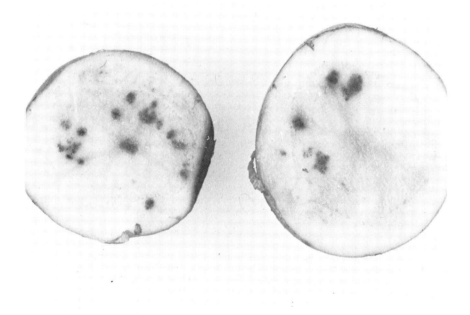

Fig. 6.8. Potato tubers cut to show internal brown spot symptoms.

different causes. Probably the most widely used name is internal brown spot. This name is applied to small, brown, round, or irregular patches of sterile tissue inside the vascular ring and usually concentrated near the center of affected tubers (Blodgett and Rich, 1950; O'Brien and Rich, 1976; Smith and Wilson, 1978). Wisconsin workers have referred to this disorder as physiological internal necrosis (Larson and Albert, 1945, 1949). Hooker (1981) called it internal heat necrosis.

Atanasoff (1926) and Conners (1967) used sprain as a synonym for internal brown spot, but Wolcott and Ellis (1956, 1959) treat sprain as a synonym for spraing or corky ring spot. Zimmermann-Griess (1947) described the same or a similar disorder of potatoes in Palestine, calling it internal rust spot. Van der Plank (1936) used the term internal brown fleck for a tuber abnormality in South Africa.

B. Cause

The exact cause of internal brown spot is not well understood. Probably several factors are involved. It appears to be most serious in large tubers produced in sandy soils under hot, dry conditions. Henninger (1979) associated it with sandy soil and high soil temperature in New Jersey. However, it can occur in muck soils. Wolcott and Ellis (1959) reported that it was more severe in 1954 under high soil moisture conditions than in 1953 under drought conditions. Wolcott (1956) associated it with conditions which favored fluctuating tuber and plant growth which favor the resorption of stored materials from the tubers. He found that fluctuating high temperatures in September were conducive to severe internal browning. Van Denburgh et al. (1979, 1980) associated brown center with cool soil temperatures (10°C) shortly after tuber initiation.

Van der Plank (1936) attributed internal brown fleck to phosphorus deficiency. Larson and Albert (1945) found that fluctuating soil temperature and moisture were more important factors than nutrition.

C. Control

Adequate soil moisture should be supplied at all times. Covering the tubers with at least 2 inches of soil or mulching with straw will decrease internal browning (Larson and Albert, 1945), but mulching with straw may be impractical.

Some cultivars are more resistant than others. Larson and Albert (1949) found that Katahdin, Ontario, Pawnee, and Russet Rural were very sus-

ceptible. Ellison and Jacob (1952) rated Green Mountain as very susceptible, Henninger (1979) classified Atlantic as highly susceptible. Bintje,
Climax, and Patrones were resistant in Lebanon (Hussain, 1965: Shuja,
1969). Invincible, Champion, Arran Banner, Kerr's Pink, and Red Skin
were susceptible, while Up-to-Date, Majestic, and Epicure were resistant in
Palestine (Zimmermann-Gries, 1947).

XVI. INTERNAL SPROUTING

A. Symptoms

Abnormal sprouts grow inward into the tuber flesh, causing cracks and
bulges. Tiny new tubers may develop inside the mother tuber (Hodgson *et
al.,* 1974).

B. Cause

Storing potatoes for long periods at high temperatures [60°F (16°C)] when
reconditioning them for chipping may induce internal sprouting of tubers.
High soil temperatures before harvest and repeated desprouting are other
factors which favor this abnormality (Hodgson *et al.,* 1974).

C. Control

Potatoes should not be stored for extended periods at high storage temperatures, and a sprout inhibitor should be used if potatoes are to be stored
beyond the normal dormant period.

XVII. JELLY END

A. Symptoms

In this disease the basal or stem ends of tubers become soft, jellylike
or translucent. This abnormality usually affects $\frac{1}{2}$–1 inch of stem end of
the tuber. Under humid storage conditions the affected tissue will retain
its shape, but under dry conditions it will dehydrate and become brown and
papery. This disorder is limited primarily to long cultivars, such as Russet
Burbank and Green Mountain, and is restricted mostly to tubers with ta-

pered or pointed stem ends (Blodgett and Rich, 1950). Other names for this disorder are jelly end rot, glassy end, glassy end rot, and translucent end (Iritani and Weller, 1973; Smith and Wilson, 1978). There are no plant symptoms associated with this malady.

B. Cause

Jelly end is the result of a marked reduction in carbohydrates in the stem end of tubers. During or following a period of stress, such as drought, the apical end of a tuber may continue to grow at the expense of the basal end, thus reducing the carbohydrate content of the basal end and also making it pointed (Iritani and Weller, 1973; Smith and Wilson, 1978).

C. Control

Selection of round cultivars will reduce the prevalence and severity of this disorder. Uniform growing conditions, including irrigation as needed, will also help to reduce the incidence of this disease (Blodgett and Rich, 1950; Smith and Wilson, 1978). Storage of mature tubers at moderate temperatures should reduce the tendency of this malady to develop during storage (Iritani and Weller, 1973).

XVIII. KNOBBINESS

A. Symptoms

One or more knobs or protuberances of various sizes and shapes, often referred to as "second growth," develop on otherwise normal primary tubers at points where eyes are located (Fig. 6.9). Occasionally tertiary tubers will develop on the secondary tubers or knobs. There are no plant or internal tuber symptoms (Blodgett and Rich, 1950; O'Brien and Rich, 1976).

B. Cause

The principal cause of knobbiness appears to be an irregular moisture supply during the growing season. The young tubers cease to enlarge at a uniform rate, and when additional moisture is supplied through rain or irrigation, the immature tubers proliferate at the eyes instead of expanding

Fig. 6.9.   Knobby potato resulting from irregular water supply. Knob at upper right shows feather and scald also due to rough handling and immaturity. (Photo courtesy of Maine Life Sciences and Agricultural Experiment Station.)

uniformly over the entire surface (Blodgett and Rich, 1950; O'Brien and Rich, 1976). Long tubers, such as Green Mountain and Russet Burbank, are especially subject to this type of malformation.

Kraus (1945) found that single-stemmed hills and poor stands favored the production of knobby tubers. Bodleander *et al.,* (1964) reported that high temperature induced second growth, irrespective of water supply.

An unusual knobby tuber disease of Katahdin has been described (Bonde and Merriam, 1957). Unlike the typical knobbiness of tubers, this is a genetically inherited abnormality (Plaisted and Peterson, 1972).

C. Control

Selection of round cultivars will reduce the knobby tuber problem. Highly susceptible cultivars, such as Green Mountain and Russet Burbank, should be avoided.

Multiple-stemmed hills and uniform stands will reduce the incidence of knobbiness. Large seed pieces or whole seed tubers produce more multiple-stemmed hills than do small seed pieces. Uniform growing conditions should be provided. This can be accomplished by controlling *Rhizoctonia* and supplying sufficient fertilizer, organic matter, and water throughout the growing season (Blodgett and Rich, 1950; Kraus, 1945; O'Brien and Rich, 1976; Rich, 1977). Avoid tuber lines of Katahdin which carry the genetic factor for knobby tubers (Plaisted and Peterson, 1972).

XIX. LIGHTNING INJURY

A. Symptoms

Dead potato plants in a more or less circular area, surrounded by healthy plants, are typical of lightning injury (Fig. 6.10). The dead area is usually 50–100 ft in diameter but it is not always circular. Usually all plants in the affected area are dead or nearly so, but an occasional plant may survive (Hooker, 1973; O'Brien and Rich, 1976).

Affected tubers have a cooked appearance (Fig. 6.11). They exhibit an extensive collapse of the pith with less injury to the cortical tissue (Hooker, 1973). Soil organisms soon invade the injured tuber tissue and cause it to break down more or less completely.

Fig. 6.10. Potato field that was struck by lightning.

Fig. 6.11.   Potato tubers from plants killed by lightning. (Photo courtesy of Maine Life Sciences and Agricultural Experiment Station.)

B. Cause

This type of injury is caused by lightning striking the field during the growing season.

C. Control

There is no practical control for this problem.

XX. SPINDLING SPROUT

A. Symptoms

Tubers produce sprouts about one-fourth the diameter of normal sprouts (Fig. 6.12). They are sometimes referred to as hair sprouts. They may be longer than normal or they may be short and weak (Blodgett and Rich, 1950; O'Brien and Rich, 1976).

Fig. 6.12. Potato tuber with spindling sprouts (top) and potato tuber with normal sprouts (bottom).

B. Cause

Spindling sprout is usually due to a lack of tuber vigor. It may be associated with leafroll, witches' broom, purple top wilt, or some other virus or mycoplasma disease, but often it occurs in the absense of any known infectious disease.

C. Control

Seed lots should be planted which are free from this disorder. Suspicious seed lots should be presprouted and tubers with spindling sprout discarded.

XXI. SPROUT TUBERS

A. Symptoms

Sometimes seed pieces will produce one or two sprouts and a new tuber will form on the end of each without ever emerging above ground or producing a plant (Fig. 6.13). Other names for this unusual disorder are "blind tubers," "secondary tuber formation," or "potatoes with no tops" (O'Brien and Rich, 1976).

B. Cause

This malady is associated with tubers which have been stored for an unusually long period of time. Warm storage temperatures also appear to favor development of sprout tubers. It tends to occur when seed pieces from old tubers are planted in cool, dry soils (Blodgett and Rich, 1950).

C. Control

Do not store seed tubers for an excessively long period of time. Store tubers at a cool temperature to prolong the dormant period and delay

Fig. 6.13.   Sprout tubers produced by tuber which produced no aboveground plant.

sprouting. Avoid planting seed pieces from old tubers in cold, dry soil. According to Conners (1967), the Chippewa, Canso, Katahdin, Ontario, and Sebago cultivars are most frequently affected. Growers may wish to select less susceptible cultivars.

XXII. STEM STREAK NECROSIS

A. Symptoms

Stem streak necrosis or manganese toxicity appears as dark brown streaks or pitted areas on the outside of stems near their base. Long, brown streaks or flecks also appear on petioles. A pale yellow-green interveinal chlorosis develops on leaves, while veins remain green. In severe cases, necrotic areas develop on leaves, followed by leaf drop, death of the terminal bud, and premature death of the plant (Berger and Gerloff, 1947).

B. Cause

The cause of stem streak necrosis, as described here, is due to manganese toxicity. Low soil pH favors manganese toxicity. This disorder should not be confused with other types of stem streaks caused by potato virus Y, *Verticillium,* or *Phytophthora infestans* (Hodgson *et al.,* 1974; Robinson *et al.,* 1960; Struckmeyer and Berger, 1950).

C. Control

Liming acid soils is the best control measure for stem streak necrosis due to manganese toxicity. The soil should be analyzed first to determine the correct amount of lime to apply because too much lime will favor scab development. Application of a balanced fertilizer containing nitrogen, phosphorus, and potash should also prove beneficial (Conners, 1967; Hodgson *et al.,* 1974).

Use of resistant cultivars will minimize losses from this disease. Cherokee, Irish Cobbler, Keswick, and Sebago should be avoided as they are highly susceptible. Canso, Green Mountain, Kennebec, McIntyre, Norgleam, and Russet Burbank (Netted Gem) are resistant (Hodgson *et al.,* 1974; Robinson and Callbeck, 1955).

XXIII. TIPBURN

A. Symptoms

Leaflets suffering from tipburn turn yellow at the tips and margins. The affected margins gradually die and turn brown or black. They roll upward and become brittle as they die and dry up (Fig. 6.14). In severe cases entire leaves may die (Lutman, 1919, 1922; O'Brien and Rich, 1976).

Fig. 6.14.   Potato leaf showing severe tipburn symptoms. (Photo courtesy of Maine Life Sciences and Agricultural Experiment Station.)

B. Cause

Tipburn is caused by excessive loss of moisture from the leaves and leaflets during hot, dry, windy weather. It begins beneath the hydathodes. Root pruning by cultivation could be a contributing factor. Sucking insects; such as leafhoppers and aphids, may accentuate the problem also (see Chapter 5, Section VII, Hopperburn).

C. Control

Irrigation during hot, dry periods is recommended. Soil organic matter and good soil tilth will improve the water-holding capacity of the soil. Effective insect control is also recommended (O'Brien and Rich, 1976).

XXIV. XYLEM RING DISCOLORATION

A. Symptoms

The vascular or xylem ring of affected tubers turns a light tan or reddish-brown color (Fig. 6.15). Discoloration is usually more pronounced at the basal or stem end of injured tubers but it may penetrate all the way to the apical end (Rich, 1950).

Fig. 6.15. Potato tubers cut to show vascular ring discoloration. This type of injury is common if potato vines are killed too rapidly. (Photo courtesy of Maine Life Sciences and Agricultural Experiment Station.)

Some tubers will exhibit a dark brown to black discoloration. However, this dark discoloration is usually incited by an infectious agent and should not be confused with the noninfectious type of xylem ring discoloration (Blodgett and Rich, 1950).

B. Cause

Xylem ring discoloration of potato tubers is caused by rapid killing of the green vines. It makes little difference if they are killed by a heavy frost or a chemical vine killer which kills the vines almost completely within 24–48 hr after application (Rich, 1950).

The darker brown or black discoloration is usually attributed to fungi, such as species of *Fusarium* or *Verticillium* (Blodgett and Rich, 1950). However, it may be caused by stem-end browning or frost (Rich, 1950).

C. Control

There is very little that can be done to control a killing frost. Chemical vine killers should be selected that kill the plants gradually over a period of several days. Two applications of half-strength chemical may produce more satisfactory results than a single application at full strength. It is important not to increase the concentration or the rate per acre beyond that recommended on the label.

7

Seed Potato Certification

Seed potato certification programs have two primary objectives. One objective is to provide seed tubers which are true to name and free from varietal mixture. This objective is rather easily attained if growers take reasonable care to keep their various cultivars separate. If a few potatoes from an extraneous cultivar become accidentally mixed with another cultivar, the shipper or grower may be able to sort them out if they have a distinctive character, such as skin color or shape. However, most cultivars cannot be identified that readily by tuber characteristics. Therefore, it is necessary to plant the tubers and rogue out the mixture, based on plant characteristics. This can be done rather easily at flowering time if the two cultivars which have become mixed have a distinctly different flower color (Fig. 7.1). Otherwise, the problem is more difficult. The roguer has to look at leaf shape, color, and other minor differences between the two cultivars. It is important that a dealer or grower can buy a certain specified cultivar with reasonable assurance that it will be true to name even though he cannot positively identify it by appearance alone. He may want a specific cultivar for a specific purpose, such as potato chips, French fries, baking, earliness, or disease resistance. If it were not for seed certification programs, one cultivar could be sold under many different names or synonyms which would result in a great deal of confusion.

The second and more important objective is the production of high-quality seed potatoes which are relatively free from disease. During the 1800s and early 1900s potato growers recognized the fact that potato cultivars degenerated or "ran out" after many years of asexual propagation. They

178

Fig. 7.1. Potato field in bloom. This is a good time to look for varietal mixture.

blamed the "running out" on continued asexual propagation per se until about 1920 when potato pathologists demonstrated that the degeneration was due to the build-up of virus diseases in the older cultivars. As the viruses are not transmitted through the true seeds, new cultivars produced originally from true seeds were revitalized until they, in turn, became infected with one or more viruses. These virus diseases came to be known as degeneration diseases.

Seed certification programs were initiated in Maine, Vermont, Idaho, and Wisconsin between 1913 and 1915, when about 5200 acres were certified (Darling, 1977). My father was a pioneer in the Maine certified seed potato industry where he started producing certified seed potatoes in 1913 or 1914 and continued to do so for the next 50 years or so. These early pioneers were ahead of their time in many ways. Very little was known about the cause and nature of the diseases which they were trying to combat. However, they soon learned how to identify leafroll, mosaics, blackleg, and other diseases by field symptoms and how to rogue them out. They obtained valuable help from federal and state plant pathologists, cooperative extension agents, and state inspectors. At other times the experts learned from the farmers. It was a cooperative venture.

Among the early potato pathologists in America were L. R. Jones, who went to Europe in 1904 to study disease resistance (Jones, 1905), and W. A. Orton, who went to Europe in 1911 and published an important bulletin based on his observations (Orton, 1914). Apparently the first United States conference on seed inspection and certification was held in Philadelphia in 1914. Orton recommended that varieties be free from mixtures. He proposed the following disease tolerances: leafroll 5%, mosaic 5%, curly dwarf 2%, blackleg 0.5%, and Fusarium and Verticillium wilts 5%, and that the total of the above diseases not exceed 5% (Darling, 1977). He recommended that the certified seed programs be administered independently at the state level and on a voluntary basis. This is the basis for the successful United States and Canadian certified seed programs as we know them today.

E. S. Schultz, a United States Department of Agriculture plant pathologist, and Donald Folsom, a Maine Agricultural Experiment Station plant pathologist, were pioneers in potato virus research in Maine. They showed that leafroll and several mosaics were caused by viruses which were transmitted by aphids (Schultz, 1919; Schultz and Folsom, 1920, 1925). They also described potato spindle tuber (1923). These men visited our farm on numerous occasions and conducted cooperative tests and demonstrations with my father. It was my good fortune to know them both personally and to study potato diseases under them. Their advice and guidance were invaluable to the Maine Department of Agriculture seed potato program and to the Maine seed potato growers.

Many states soon developed certified seed potato programs (Figs. 7.2–7.4). In most cases they were supervised by the State Department of Agriculture. In other instances they were supervised by the Cooperative Extension Service, a state seed board or commission, or a seed growers' organization. Canada had one organization for the entire dominion.

It soon became evident that the northern states and Canadian provinces were best adapted to seed production. It was thought that northern grown potatoes grown in a cool climate retained more vigor than potatoes grown in a warm climate. Probably a more important factor was the lower population of aphids in the northern latitudes and higher elevations which resulted in less spread of virus diseases. For the same reasons Scotland has proven to be a very good seed-producing area in the British Isles.

Nineteen states in the United States now produce almost 200,000 acres (80,000 ha) (Darling, 1977) and Canada produces about 65,000 acres (26,000 ha) (Munro, 1975) of seed potatoes. Ninety different cultivars were represented in 1975 (Turnquist, 1976). Much of Canada's production is exported. The leading cultivars in the United States, in order of popularity are Russet Burbank, Kennebec, Katahdin, Superior, and Norchip with over 10,000 acres each certified in 1975. Other cultivars with more than 1000

Fig. 7.2. Field of Irish Cobbler certified seed potatoes grown in isolated area of northern New Hampshire.

Fig. 7.3. Field of Green Mountain certified seed potatoes in high altitude, isolated area of northern New Hampshire.

Fig. 7.4. Potato inspector inspecting field of Green Mountain potatoes entered for certification.

acres include Norgold Russet, Norland, Red Pontiac, Red La Soda, Monona, Irish Cobbler, White Rose, La Rouge, Sebago, and Chippewa (Darling, 1977). Leading seed potato cultivars in Canada are Kennebec, Sebago, Katahdin, Red Pontiac, Netted Gem (Russet Burbank), Green Mountain, and Irish Cobbler (Munro, 1975).

Most certified seed lines (clones) start from a single tuber or single hill. A tuber is indexed by removing one eye and planting it. A hill is indexed by planting one tuber or one eye from a tuber representing that hill. If the plant shows any symptoms of disease or if indicator plants show any symptoms when inoculated with sap from the index plant, the original tuber or hill is discarded.

Considerable progress has been made in recent years in producing virus-free clones via meristem culture and heat therapy. Most of the old cultivars such as Russet Burbank and Green Mountain were 100% infected with potato virus X. Some were also infected with other symptomless viruses such as potato virus S. Clones of these cultivars have been freed from all recognizable viruses via such techniques as meristem culture and/or heat therapy. The presence or absence of symptomless viruses is determined by assaying on sensitive indicator plants or by serology.

The next problem is how to maintain these new virus-free clones free from reinfection. This is accomplished in various ways. Several states and provinces plant these virus-free clones by the tuber-unit method on isolated foundation seed farms. The tuber-unit method consists of cutting each potato in pieces, usually four, and planting the pieces contiguously in the row. During the growing season each unit is examined carefully. If there is any evidence of disease or other abnormality the entire unit is removed (rogued). A close watch is kept for the presence of aphids which spread the more important viruses such as leafroll virus, PVA, and PVY. If any aphids are observed, plants are sprayed with an aphicide. Spraying may be done by airplane (Fig. 7.5) or helicopter (Figs. 7.6 and 7.7) to avoid the risk of spreading sap-transmissible viruses such as PVM, PVS, and PVX by the tractor or sprayer wheels. This also applies to the potato spindle tuber viroid. Vines are often killed in late summer or early fall to avoid the risk of late-season infestations of aphids which might introduce and/or spread viruses (Fig. 7.8).

Seed certification in Canada follows a definite pattern. Virus-free clones

Fig. 7.5. Spraying potato field by airplane for insect and disease control.

Fig. 7.6. Potato field in Aroostook County, Maine, being sprayed by helicopter for insect and disease control.

Fig. 7.7. Helicopter application of chemicals to potato field. Note swirling pattern of concentrated spray mist.

Fig. 7.8. Chemical vine killing to prevent late season spread of virus diseases and late blight. The row between the two men was mistakenly left untreated.

are planted and the tubers are harvested to produce pre-elite bulked clones. These tubers are planted the following year to produce Elite I seed. The Elite I seed is increased to produce Elite II seed. Elite II seed is used to produce Elite III seed which, in turn, is used to produce foundation seed, and foundation seed is planted to produce certified seed (Munro, 1975, 1978). Certified seed cannot be planted for recertification.

Some states in the United States have a similar progression, but most of them are less regimented. More emphasis is usually placed on a winter test program. Samples of each lot are planted in Florida, California, or another southern state or in a greenhouse. Trained inspectors examine the growing plants for presence of absence of viruses. Seed lots must pass a rigid inspection with very little virus present to be classified as foundation seed. Winter test readings should represent a rather accurate indication of the virus content of each seed lot.

In addition to viruses, viroids, and mycoplasmas, attention is also paid to bacterial ring rot, blackleg, and wilts (Verticillium and Fusarium) in the production of certified seed potatoes. Special emphasis is placed on bacterial ring rot. This disease is so contagious that it has a zero tolerance.

In order for potatoes to be certified in Maine, they must pass two field inspections within the following tolerances (see tabulation):

	First inspection (%)	Second inspection (%)
Leafroll	2	1
Mosaics	3	2
Spindle tuber	2	2
Total virus and viroid	5	3
Varietal mixture	1	0.25
Bacterial ring rot	0	0

In addition, Chippewa, Katahdin, Norchip, Ontario, Russet Burbank, and Superior must pass a Florida test with less than 5% virus. In order to qualify for foundation seed, the seed lot must be tested in Florida and not show more than 0.5% leafroll, mosaics, and spindle tuber combined (Maine Seed Potatoes Certified, 1978).

Seed certification in Idaho is the responsibility of the Idaho Crop Improvement Association (1978). The tolerances are given in the tabulation below.

	First inspection (%)	Second inspection (%)
Varietal mixture	0.5	0.2
Weak plants	10.0	5.0
Mosaic, leafroll, etc.	2.0	1.0
Leafroll (singly)	0.2	0.2
Blackleg	2.0	2.0
Ring rot	0.0	0.0
Eumartii wilt	0.0	0.0

Each lot must be tested in the winter test program and must not exceed 2% leafroll. In addition a voluntary potato virus X test program is available to seed growers. Most seed potatoes in Idaho are Russet Burbanks.

Requirements for Maine and Idaho are given here only as examples, and were selected because they are two of the leading seed-producing states. Tolerances may change slightly from year to year, but they tend to remain relatively constant. Most other states have similar requirements for field inspections, winter test programs, and tolerances.

Glossary*

Å (Ångstrom)—A unit of length equal to 1 nanometer (nm)

Abaxial—Referring to upper leaf surface. Cf. Adaxial

Abiotic—Nonliving, nonparasitic. Cf. Biotic

Absciss—Shedding of leaves due to cambial activity of cells at the base of the petiole

Acervulus—A subepidermal, saucer-shaped, asexual fruiting body bearing short conidiophores and conidia

Acid—Sour; having a pH below 7.0. Cf. Alkaline

Acre—A unit of land measurement equal to 43,560 ft^2 or 0.405 ha

Actinomycete—A group of microorganisms intermediate between bacteria and fungi; includes *Streptomyces scabies,* the causal agent of common scab

Adaxial—Referring to lower leaf surface. Cf. Abaxial

Adsorb—Hold on a surface

Aerobic—Requiring free oxygen for growth. Cf. Anaerobic

Agar—A gelatin-like material obtained from seaweed and used to prepare media for growing microorganisms

Agglutination—A serological test in which viruses or bacteria suspended in a liquid clump together when the suspension is treated with antiserum containing antibodies specific against these viruses or bacteria

Alkaline—Basic; having a pH above 7.0. Cf. Acid

Alkaloid—An organic compound with alkaline properties, usually poisonous, present in potato foliage and green tubers

Alternaria—A fungus genus in the Moniliales, causal agent of early blight

Anaerobic—Not requiring free oxygen for growth. Cf. Aerobic

Anastomosis—Union of one organ (for example, hypha) with another, resulting in an exchange of contents

*Note: To make the plural of Latin words ending in:
-*us,* change *us* to *i* (e.g., ascus, asci); -*um,* change *um* to *a* (e.g., conidium, conidia); -*a,* change *a* to *ae* (e.g., hypha, hyphae); -*is,* change *is* to *es* (e.g., anastomosis, anastomoses).

187

Antheridium—The male sexual organ of some fungi, for example, the Oomycetes

Anthocyanin—A reddish or purple pigment in plant parts such as flowers, leaves, and stems

Antibiosis—An association between two organisms which is harmful to one of them

Antibiotic—A chemical compound produced by one microorganism which inhibits or kills other microorganisms

Antibody—A protein produced in an animal in response to an injected foreign antigen

Antigen—A substance, usually a foreign protein, capable of stimulating the formation of antibodies

Antiserum—The blood serum of a warm-blooded animal that contains antibodies

Apex—The tip of a plant organ; that portion of a root, shoot, or tuber containing apical meristem

Aphid (or aphis)—A plant louse belonging to the Hemiptera

Apical—At the end of (apex)

Apical end—The bud, eye, or rose end of a tuber. Cf. Stem end

Apothecium—An open cup- or saucer-shaped ascocarp of the Discomycetes, a subdivision of the Ascomycetes

Appressed—Lying flat or pressed closely against something, as leaves on a stem

Appressorium—An organ of attachment or penetration produced by hyphae or germ tubes of certain fungi

Ascocarp—The fruiting body of Ascomycetes containing asci

Ascogonium—The female gametangium or sexual organ of Ascomycetes

Ascomycetes—A group of fungi, producing their sexual spores, ascospores, within asci

Ascospore—A sexual spore borne in an ascus

Ascus—A saclike structure containing ascospores (usually 8)

Asexual reproduction—Reproduction not involving the union of gametes (for example, reproduction by conidia)

Avirulent—Lacking ability to be pathogenic (weak)

Axillary bud—A bud formed in the upper angle between a stem and petiole

Bacillus—A rod-shaped bacterium

Bactericide—A chemical that kills bacteria

Bacteriophage (phage)—A tiny virus that infects bacteria

Bacteriostatic—Inhibits bacterial growth but does not kill bacteria

Bacterium—Small (0.5 – 2.0 μm) unicellular organism that lacks chlorophyll and multiplies by fission

Basidiomycetes—A class of fungi producing their sexual spores, basidiospores, on basidia

Basidiospore—A sexually produced spore borne on a basidium

Basidium—A club-shaped structure on which basidiospores are borne

Biflagellate—Having two flagella

Biotic—Caused by another living agent. Cf. Abiotic

Biotype—A population of individuals having identical genetic characters; a subdivision of a pathogenic race

Black dot—A weakly parasitic disease of potato caused by *Colletotrichum atramentarium* (Beck. & Br.) Taub.

Blackleg—A disease caused by *Erwinia atroseptica*

Black scurf—A disease caused by *Rhizoctonia solani* Kühn

Blight—A disease characterized by general and rapid killing of leaves, flowers, and stems, e.g., early blight, late blight

Blotch—A disease symptom characterized by large, irregular, spots or blots on leaves, stems, or shoots and often accompanied by fungus growth

Bud—Unopened blossom; eye on potato tuber

Bud end—Apical end or eye end of tuber. Cf. Stem end

Bushel—35.24 liters; bushel of potatoes weighs 60 lbs (27.24 kg)

Callus—A mass of thin-walled cells, usually developed as a result of wounding or cutting potatoes, sometimes called cork

Cambium—A thin layer of meristematic cells. It produces all secondary tissues and results in growth in diameter

Canker—A necrotic, usually sunken lesion on a stem or tuber, e.g., Rhizoctonia canker on underground stem

Causal agent of disease—Capable of causing disease, e.g., bacterium, fungus, virus

Celsius—A temperature scale based on the freezing point of water at 0° and the boiling point at 100°. Also called centigrade. One degree Celsius equals 5/9° Fahrenheit. To convert Celsius to Fahrenheit, multiply by 1.8 and add 32

Centimeter—About 0.4 inch

Certified seed—Seed that has been inspected for varietal purity and freedom from disease within set limits

Chemotherapy—Treatment of a plant disease with chemicals that are absorbed and are translocated internally

Chit—To green sprout seed potatoes

Chlorophyll—The green coloring matter of plants

Chloroplast—A plastid containing chlorophyll

Chlorosis—Yellowing of normally green tissue due to lack of development of chlorophyll

Chromosome—A filamentous structure in the cell nucleus, along which the genes are located

Chronic symptoms—Symptoms that remain over a long period of time or recur year after year

Class—One of the main subdivisions of a phylum

Cleistothecium—A closed ascocarp, e.g., powdery mildew

Clonal propagation—Asexual propagation from tubers, cuttings, etc.

Clone—A group of organisms descended asexually from a single common ancestor, hence genetically identical

Coenocytic—Nonseptate, multinucleate, as applied to mycelium in the Phycomycetes

Concentric—Forming one circle around another with a common center, e.g., early blight lesions on leaves

Conidiophore—A specialized hypha on which one or more conidia are produced

Conidium—An asexual fungus spore formed at the end of a conidiophore

Control—Reduction in crop losses; also applied to untreated "check" or standard when used to compare with an experimental treatment

Cortex—The stem or root tissue between the epidermis and the phloem

Corynebacterium—A genus of bacteria; *C. sepedonicum* is the causal agent of ring rot

Cotyledon—The seed leaf; one in the monocotyledons, two in the dicotyledons

Cross protection—Plant tissues infected with one strain of a virus are protected from infection by other strains of the same virus

Cryptogam—A plant that produces no flowers or seeds but is propagated by spores, e.g., fungi.

Cryptogram—Something written in code, e.g., characteristics of viruses

Cull—A potato which is unmarketable due to size, shape, injury, or disease

Cultivar—A horticulturally derived variety of a plant as distinct from a botanical variety

Culture—The growing of microorganisms in a nutrient medium, also tillage of the soil

Culture medium—The prepared food material on which microorganisms are grown

Cuticle—The outer wall of epidermal cells of a leaf consisting primarily of wax and cutin

Cyst—A saclike structure; in fungi, an encysted zoospore; in nematodes, the carcass of a dead adult female in the genus *Globodera*, which usually contains eggs

Cytoplasm—The living substance of a cell outside the nucleus

Decay—To decompose, rot

Deoxyribose nucleic acid—See DNA

Desiccant—A chemical used to hasten drying; a vine killer

Desiccate—To dry up

Dieback—Progressive death of shoots, starting at the tips

Diploid—A plant having twice the haploid number of chromosomes

Discomycete—An ascomycete whose fruiting body is an apothecium

Disease—Injurious physiological processes caused by a continuous irritation of a plant by a primary causal agent, resulting in pathological symptoms

Disinfect—To inactivate or remove a pathogen from its host

Disinfectant—A physical or chemical agent that frees a plant, organ, or tissue from infection

Disinfest—To kill pathogens in or on soil, tools, seed surface, etc., before they initiate disease

Disinfestant—An agent that kills or inactivates pathogens in the environment or on the surface of a plant organ before infection takes place

Dissemination—Transfer of inoculum to healthy plants

Distribution—Long-distance spread of a pathogen to areas outside its previous geographical range.

DNA—Deoxyribonucleic acid, the genetic material of organisms, animal viruses, and a few plant viruses. It is found in mycoplasmas also

Dormant—Inactive, applied to potato tubers which have not sprouted

Ectoparasite—Parasite living on the outside of its host. Cf. Endoparasite

Edema—Intumuscences or blisters on leaves due to intercellular water pressure

Eelworms—Nematodes

Enation—An eruption or outgrowth of tissue from a plant part

Encyst—To form a cyst

Endemic—Occurring naturally in a country or area

Endodermis—Plant tissue surrounding the vascular cylinder

Endogenous—Developing from within. Cf. Exogenous

Endoparasite—Parasite living within its host. Cf. Ectoparasite

Enphytotic—Damage from a plant disease that remains about constant from year to year. Cf. Epiphytotic

Enzyme—A protein produced by living cells that can catalyze a specific organic reaction

Epidemic—Severe outbreak of disease. Cf. epiphytotic

Epidemiology—The study of epidemics. Cf. epiphytology

Epinasty—Downward curling of a leaf blade

Epiphytology—The study of epiphytotics

Epiphytotic—Severe outbreak of a plant disease

Eradicant—A chemical substance that destroys a pathogen

Eradication—Destruction or removal of a pathogen already established in a given area

Erysiphe—A genus of powdery mildew

Escape—Lack of disease development, even though the pathogen is prevalent

Etiolation—Elongation of internodes and yellowing of tissue due to lack of light

Etiology—The study of the causal agent and its relationship to host plants

Exclusion—Keeping a pathogen out of a disease-free area

Exogenous—Developing on the outer surface. Cf. Endogenous

Extracellular—Outside the cells. Cf. Intracellular

Exudate—Liquid discharge from diseased or healthy plant tissue

Eyes—The buds on potato tubers from which sprouts arise

Facultative parasite—An organism that is usually saprophytic but which can become parasitic under certain conditions. Cf. Facultative saprophyte; Obligate parasite

Facultative saprophyte—An organism that is usually parasitic but can live saprophytically. Cf. Facultative parasite.

Fahrenheit—A temperature scale based on the freezing point of water at 32°F and the boiling point of water at 212°F. To convert °F Fahrenheit to Celsius, subtract 32 and divide by 1.8. Cf. Celsius

Fasciation—Flattening of stems and/or tubers

Fasciculation—Excessive shoot development producing a witches-broom effect

Feather—Loose skin of potato tuber. Cf. Scald

Field-immune—Plants which, when infected, react in a hypersensitive fashion, i.e., the pathogen is localized at the point of entry and the establishment of the disease prevented

Field-resistant—Plants which limit the rate of infection to such an extent that the eventual damage is significantly less than the average of cultivars

Filamentous—Threadlike, filiform

Filiform—Threadlike

Fission—Transverse splitting or dividing of bacterial cells

Flagellum (pl. flagella)—A whiplike appendage of a bacterium or zoospore, used for locomotion

Flagging—Loss of turgidity and drooping of leaves just prior to wilting of a plant

Fleck—A small spot

Foot—About 0.3 meter

Forma specialis **(Abbr. F. sp.)**—Special form; a biotype of a species of pathogen that differs from others in its ability to infect. Cf. Physiologic race

Fruiting body—A structure that bears fungus spores; examples: acervuli, apothecia, perithecia, pycnidia

Fumigant—A toxic gas that kills organisms. Usually used in soils or closed structures, often applied as a volatile liquid

Fumigate—To apply a volatile chemical to soil or use it in a closed structure to kill pathogens

Fungicide—A chemical that is toxic to fungi

Fungi Imperfecti—A form class containing those fungi which lack a sexual stage or whose sexual stage is unknown

Fungus—A thallus plant (lacking roots, stems, and leaves) which lacks chlorophyll. Its vegetative body consists of hyphae.

Fusarium—A genus in the Fungi Imperfecti

Gall—An overgrowth, tumefaction, or tumor

Gallon—3.785 liters

Gametangium—The cell or organ in which gametes develop

Gamete—A male or female reproductive cell or the nuclei in a gametangium

Gangrene—Death and decay of tissues

Gel—A jellylike colloidal mass

Gene—A tiny functional hereditary unit that occupies a fixed position on a chromosome

Genome—A complete haploid set of chromosomes

Genotype—The particular combination of genes present in the cells of an individual

Genus—A taxonomic category ranking below a family and above a species

Germination—The beginning of growth of a spore or seed

Germ tube—The hypha produced by a fungal spore when it begins to grow

Giant hill—A large, rough oversized potato plant or plants in a hill which produce large, rough potatoes. The exact cause is not known

Gram—1/1000 kilogram; 0.0353 ounce

Gram negative—Bacteria that retain red stain from gram stain

Gram positive—Bacteria that retain blue stain from gram stain

Gram stain—A differential bacterial stain developed by Gram

Greening—Green pigmentation of potato tuber skin and flesh due to exposure to light

Habitat—The natural location where an organism occurs

Haploid—Having a single set of unpaired chromosomes in each nucleus

Haulm—The stem or stalk of a potato plant

Haustorium—Fingerlike absorbing hyphae which invade a host cell

Hectare—Land measurement equal to 2.47 acres (Abbr. = ha)

Heel end—Stem end of potato tuber

Herbaceous plant—A higher plant which is nonwoody

Herbicide—A chemical used to kill weeds

Hill—A plant or group of plants produced by one seed piece

Hollow heart—Potato tuber with a cavity in its center

Host—A living organism from which a pathogen derives its sustenance

Host range—The various kinds of host plants that are susceptible to attack by a given pathogen

Hyaline—Colorless, transparent

Hybrid—The offspring of genetically dissimilar parents or stock

Hybridize—Crossbreed

Hydathode—Opening in leaf through which water is released

Hydrosis—Symptom characterized by water-soaked tissue

Hyperplasia—Plant overgrowth resulting from abnormal cell division. Cf. Hypertrophy

Hypersensitivity—Very sensitive reaction of plant tissues to certain pathogens. Affected cells are killed rapidly, thus preventing further spread of the pathogen

Hypertrophy—Plant overgrowth resulting from abnormal cell enlargement. Cf. Hyperplasia

Hypha (Pl. hyphae)—A single mycelial thread

Immune—Completely resistant. Cf. Resistant; Susceptibility; Tolerant

Imperfect stage—Sexual spores are lacking or unknown. Cf. Perfect stage

Intracellular—Inside the cell. Cf. Extracellular

Isolate—A single spore or culture and its subcultures; to separate a pathogen from its host and culture it

Isolation—Separation from other plants or fields

Juvenile—Young, immature

Kilogram—1000 grams; 2.205 pounds

Kilometer—1000 meters; 0.621 mile

Knobbiness—Second growth or proliferation of more or less healthy tissue protruding from the surface of tubers

Larva —An insect in its earliest stage of development; an immature nematode

Latent infection—An infected plant which shows no symptoms

Latent virus—A virus which produces no symptoms in its host

Leafhopper—A sucking insect in the family Cicadellidae

Leafroll—A virus disease of potato; leaf curl

Leaf spot—A small, self-limiting lesion on a leaf

Leak—A watery soft rot of potato tubers

Lenticel—A respiratory pore in a stem of a plant; a breathing pore on the surface of a potato tuber

Lesion—A localized area of diseased tissue

Life cycle—The succession of changes which an organism undergoes during its life

Lift—To dig potato tubers out of the soil (British)

Liter—1.057 quart; 1000 ml

Local infection—Infection confined to a small part of the plant

Local lesion—A localized spot produced on a leaf of an indicator plant after inoculation with a virus

Macroscopic—Large; visible without the aid of a lens. Cf. Microscopic

Maneb—Manganese ethylene bisdithiocarbamate fungicide.

Masked symptoms—Absence of symptoms under certain environmental conditions even though the plant is diseased

Mechanical inoculation—Sap from a virus-infected plant is rubbed on a healthy plant

Medulla—Central part of an organ, e.g., center of a potato tuber

Meristem—The growing point or area of rapidly dividing cells at the tip of a stem, root, or branch

Mesophyll—Parenchyma cells between epidermal layers of leaves

Meter—3.281 feet or 1.094 yards

Microbe—A microscopic organism.

Micron—A unit of length equal to 1/1000 millimeter.

Microorganism—A microscopic organism, e.g., bacterium, fungus. Cf. Microbe

Microscopic—Very small, visible with a lens only. Cf. Macroscopic

Mildew—Downy or powdery fungal growth on the surface of a diseased plant

Mile—About 1.6 kilometers

Millimeter (mm)—A unit of length equal to 0.1 centimeter; about 0.04 inch

Millimicron (mμ)—A unit of length equal to 0.001 micron or 1 nanometer

Mold—Profuse superficial fungal growth on a substrate

Molt—The shedding of a cuticle, e.g., nematodes

Monokaryotic—Containing one nucleus

Mosaic—Symptom of some virus diseases characterized by mixed patches of normal and light green or yellow color

Motile—Capable of spontaneous movement

Mottle—An irregular pattern of light and dark areas on leaves

Mummify—To shrivel or dry up

Mummy—Usually applied to a dried, shriveled fruit but sometimes applied to dried, shriveled potato tuber

Mushroom—A fleshy fruiting body of a fungus in the Basidiomycetes

Mutation—A sudden, permanent genetic change

Mycelium—Mass of hyphae that constitute the thallus or vegetative body of a fungus

Mycology—The study of fungi

Mycoplasma—Microscopic organisms intermediate in size between viruses and bacteria. They have a single unit membrane, contain both DNA and RNA, and are sensitive to tetracycline

Mycotic—Caused by fungi

Mycorrhiza—A symbiotic association of a fungus with the roots of a plant

Myxomycetes—The slime molds, a low class of fungi characterized by amoeboid vegetative protoplasts and plasmodia

Nabam—Disodium ethylene bisdithiocarbamate

Nanometer (nm)—10^{-9} meter; 1 millimicron

Natural openings—Hydathodes, lenticels, nectaries, stomata

Necrosis—Death of plant cells

Necrotic—Dead and discolored

Nema—Nematode, eelworm

Nematicide—A chemical that kills nematodes

Net necrosis—First-year symptom of tuber infection with leafroll virus; phloem necrosis of tubers

Noninfectious disease—A disease caused by an environmental factor, not by a parasite. Cf. Abiotic; Nonparasitic disease

Nonparasitic disease—An abiotic or noninfectious disease

Nonseptate—Coenocytic; lacking cross walls, applied to mycelium in the Phycomycetes

Nucleic acid—A high-molecular-weight acid that contains ribose or deoxyribose. The infectious parts of viruses contain nucleic acids

Nucleoprotein—Chemical consisting of nucleic acid and protein

Nucleus—The dense protoplasmic body in each cell

Nymph—A young insect that undergoes incomplete metamorphosis, often applied to immature leafhoppers

Obligate parasite—An organism that in nature can live and grow on another living organism. Cf. Facultative parasite; Facultative saprophyte

Oogonium—Female reproductive structure of Oomycetes, e.g., *Phytophthora, Pythium*

Oomycetes—Phycomycetes which produce oospores

Oospore—A sexual spore formed by the fertilization of an oogonium by an antheridium

Order—A subdivision of a class, itself divided into families

Organic compound—A chemical containing carbon

Osmosis—The fusion of a solvent through a semipermeable membrane

Ostiole—A pore; an opening in a fruiting body through which spores are discharged

Oxidation—The combination of a substance with oxygen

Oxygen (O_2)—A gas essential for plant and animal life

Ozone (O_3)—A highly reactive form of oxygen that is injurious to plants if present in more than trace amounts

Parasite—An organism living on or in another living organism and obtaining its food from the latter (host). Viruses are usually considered to be parasites as well as parasitic bacteria, fungi, and nematodes

Parenchyma—Plant tissue composed of thin-walled, loosely packed, relatively unspecialized cells

Pathogen—An agent which can cause disease

Pathogenesis—That stage of the life cycle of a pathogen during which it is associated with its host

Pathogenic race—See physiologic race

Pathogenicity—The relative capability of a pathogen to cause disease

Pathology—The study of disease

Perfect stage—That part of the life cycle of a fungus during which sexual spores are produced. Cf. Imperfect stage

Pericycle—A layer of cells inside the endodermis but outside the phloem of roots and stems

Periderm—The corky outer bark of older stems and leaves

Perithecium—Flask-shaped fruiting body in the Ascomycetes, having an ostiole

Permeable—Permitting other substances to pass through

Petiole—The stalk of a leaf

Phenotype—The physical manifestation of a genetic trait

Phloem—Vascular tissue, consisting of sieve tubes, companion cells, phloem parenchyma, and fibers, which conduct food

Photosynthesis—Autotrophic synthesis of organic chemicals in which the source of energy is light

Phycomycetes—A class of fungi having nonseptate mycelium.

Phyllody—Production of leaves instead of petals in flowers

Physiologic race—A biotype, or group of closely related biotypes, alike in morphology, but unlike other biotypes in certain physiological or pathological characters

Physiology—The life processes and functions of organisms

Phytophthora—A genus of fungi in the Phycomycetes. *Phytophthora infestans* is the causal agent of late blight

Phytotoxic—Injurious to plants

Pith—A tissue (usually parenchyma) located in the center of a stem or tuber (rarely a root), internal to the xylem

Plasmodium—A naked mass of protoplasm containing numerous nuclei

Polyhedron—A spheroidal particle or crystal with many plane faces

Precipitin—An antibody that precipitates soluble antigens

Predisposition—An increase in susceptibility due to the influence of the environment upon the suscept

Primary infection—First infection; start of a new life cycle following a dormant period

Proboscis—A slender, tubular feeding and sucking structure of some insects; stylet

Proliferation—A rapid and repeated production of new cells, tissues, or organs

Propagative virus—A virus that multiplies in its insect vector

Propagule—A structure capable of giving rise to a new individual

Protectant—A substance that protects a plant against infection

Protection—A principle of plant disease control; placing a barrier (e.g., chemical spray or dust) between suscept and pathogen

Psylla or *Psyllid*—A sucking insect in the family Psyllidae

Pupa—A stage of some insects between the larva and adult

Purification—The separation of virus particles in a pure form

Pycnidium—A flask-shaped fungal fruiting body containing conidia

Quarantine—Control of import and export of plants to prevent spread of disease and pests

Race—See physiologic race

Range—The geographical region in which a plant pathogen is known to occur

Recessive —An heritable character or gene which is expressed only when present in a homozygous condition

Resistance—The ability of a plant to overcome, to some degree, the effect of a pathogen

Resistant—Possessing characteristics that hinder or retard the development of a given pathogen. Cf. Immune; Susceptibility; Tolerant

Resting stage—A fungal spore, usually thick-walled, that can remain viable in a dormant condition for an extended period

Resting spore—An inactive stage of a fungus, usually a thick-walled spore

Reticulate—Covered with netlike ridges

Rhizome—A modified stem growing horizontally along or under the ground. Cf. Stolon

Rhizomorph—A compact strand of hyphae

Rhizosphere—The soil near a living root

Ribonucleic acid—See RNA

Ring rot—Bacterial ring rot caused by *Corynebacterium sepedonicum*

Ringspot—A circular area of chlorosis with a green center; a symptom of certain virus diseases

RNA—Ribonucleic acid; the only nucleic acid of most plant viruses

Rogue—Remove undesirable plants, especially those infected by a virus, from the growing crop

Rose end—Apical or eye end of a potato tuber

Rosette—Short, bunchy habit of plant growth

Rot—To undergo decomposition; decay

Roundworm—Nematode

Rugose—Rough; rough or crinkled leaves infected with a virus, e.g., rugose mosaic

Russet—Rough, brownish area on the skin of a potato tuber as a result of cork cell formation

Rust—A disease caused by a rust fungus (order Uredinales); also the fungus itself

Sanitation—Decontamination of tools, equipment, etc. (Cf. Disinfestant); also, removal and destruction of infected plant parts

Saprophyte—An organism that lives on dead organic matter

Scab—A hyperplastic symptom characterized by rough, crusty lesions

Scald—Appearing burned, as if with hot water or steam

Sclerotium—A compact mass of hyphae in a dormant state

Scorch—Burning of leaves or leaf margins due to infection or unfavorable environmental conditions

Secondary infection—Infection caused by inoculum from a primary cycle or another secondary cycle. Cf. Primary infection

Secondary inoculum—Inoculum produced by infections during the same growing season

Seed-piece—Portion of a seed potato tuber used for propagation

Septate—Having cross walls or septations

Serology—Use of the antigen–antibody reaction for the detection and identification of antigenic substances and the organisms that carry them

Set—Sometimes used in place of seed-piece; the number of potato tubers in a hill (e.g., a good set)

Sexual—Union of nuclei in which meiosis takes place

Shock symptoms—Severe, often necrotic symptoms on the first new growth after infection with certain viruses

Shot-hole—A symptom in which small diseased fragments of leaves fall out and leave numerous small holes in the leaf blade

Shrivel—To lose turgidity; to become flabby

Signs—Visible structures of a pathogen produced in or on diseased tissues (e.g., sclerotia, spores)

Smut—A disease caused by smut fungi (order Ustilaginales); also the fungus itself

Species—Subdivision of a genus. A group of closely related individuals morphologically distinct from other individuals

Spindle tuber—A disease caused by the spindle tuber viroid

Spiroplasma—Similar to mycoplasma but spiral-shaped

Sporangiophore—A specialized hypha bearing one or more sporangia

Sporangiospore—Nonmotile, asexual spore borne in a sporangium

Sporangium—A more or less spherical body in which asexual spores are produced

Spore—A reproductive structure in the fungi

Sporulate—To produce spores

Spot—A symptom of disease characterized by small necrotic areas

Spraing—A Scottish dialect word meaning "streak"; applied to necrotic rings or arcs in potato tubers infected with certain soil-borne viruses such as tobacco rattle virus, also called corky ring spot

Sprout—A shoot arising from a bud or an eye of a potato tuber

Stalk—Haulm; stem of a plant

Stem end—Basal end of tuber; point of stolon attachment

Stem-end browning—A dark brown discoloration of xylem and phloem occurring at the stem end of potato tubers

Sterile—Applied to fungi which lack spores. Also free from living organisms

Stolon—In potato, an underground stem which produces a tuber at its terminus

Stoma—An opening, regulated by guard cells, in the epidermis of a leaf or other plant part

Streptomyces—A genus in the Actinomycetales. *Streptomyces scabies* is the causal agent of common scab of potato

Stroma—A mass of vegetative hyphae on or in which fungus fruiting bodies are usually formed

Stylet—A long, slender, hollow feeding structure of nematodes and some insects

Stylet-borne—A virus borne on the stylet of its vector; a noncirculative virus

Suberin—A waxy material found in the walls of cork cells

Suberized—Refers to cell walls hardened by their conversion to cork (suberin)

Subspecies—A genetically distinctive geographic subunit of a species

Suscept—Any plant that can be attacked by a given pathogen; a host plant or potential host plant

Susceptibility—The inability of a plant to resist or avoid disease

Symbiosis—A mutually beneficial association of two different kinds of organisms

Symptom—A visible expression of a pathological condition or disease

Syndrome—A number of symptoms occurring together and characterizing a specific disease

Synonym—Another name for a species, either for a plant or a pathogen

Systemic—Spreading internally throughout the plant; may be applied to a chemical or a pathogen

Taxonomy—The classification of organisms based on their evolutionary relationships

Terraclor—Pentachloronitrobenzene

Thallus—A plant body lacking true roots, stems, and leaves

Tolerance—The ability to endure disease without serious injury or crop loss

Tolerant—Showing little reaction to infection by a pathogen. Cf. Resistant; Susceptibility

Translocation—Transfer of a chemical or virus through the plant

Transmission—Dissemination of pathogens and inoculation of suscepts

Transpiration—The loss of water vapor from the surface of leaves

Tuber—A short, fleshy underground stem, such as the potato, bearing buds from which new plant shoots arise

Tuber-perpetuated—Propagated asexually by a tuber or portion of a tuber; also any pathogen carried through the tuber to succeeding generations

Tuber unit—Planting the pieces of a tuber in sequence

Tumefaction—A plant tumor or gall

Uniflagellate—Possessing one flagellum

Vapam—Sodium methyl dithiocarbamate, a soil fumigant requiring a water seal

Variety—Botanically, a subdivision of a species. Often used incorrectly in place of cultivar. Cf. Cultivar

Vascular—Applied to conductive tissue of plants

Vector—An agent of dissemination of inoculum, usually applied to animals such as insects, nematodes, etc.

Vegetative reproduction—Asexual reproduction

Vein banding—Retention of bands of green tissue along the veins while the tissue between the veins becomes chlorotic

Vein clearing—A symptom of virus-infected leaves where veinal tissue is lighter green than normal

Vessel—A xylem element or elements whose function is to conduct water and mineral nutrients

Viable—Able to live; spores which are able to germinate

Virescence—A symptom in which green pigment develops where it does not normally occur, e.g., greening of potato tubers

Virion—A complete virus particle

Viroid—A subviral agent containing RNA but lacking a protein coat, Cf. spindle tuber

Virulence—The degree of pathogenicity of a given pathogen

Virulent—Strongly pathogenic

Viruliferous—Term applied to a vector which contains a virus and is capable of transmitting it

Virus—A submicroscopic (one dimension less than 200 nm), obligate parasite consisting of a nucleic acid core surrounded by a protein coat

Vorlex—Methyl isothiocyanate-dichloropropene mixture; a soil fumigant for control of nematodes, fungi, and other soil organisms. No seal is usually required

Wilt—A plant symptom characterized by loss of turgor, resulting in drooping of leaves and other above-ground plant parts; a disease whose principal symptom is wilting

Witches'-broom—A hyperplastic symptom resulting from the abnormal development of many weak shoots from adventitious buds

Xylem—Vascular tissue consisting of tracheids, vessels, fibers, and parenchyma cells which transport water and dissolved substances upward in the plant; woody tissue of a plant

Yellows—A plant disease characterized by yellowing and stunting of the host plant

Zoosporangium—A sporangium that produces zoospores

Zoospore—A motile, asexual spore produced in a sporangium; swarmspore

Zygospore—A thick-walled resting spore formed by the union of two morphologically similar gametangia

Zygote—A diploid cell formed by the union of two gametes

Bibliography

Abdel-Rahman, T. B., Elgindi, D. M., and Oteifa, B. A. (1974). Efficacy of certain systemic pesticides in the control of root-knot and reniform nematodes of potato. *Plant Dis. Rep.* **58**, 517–520.

Agrios, G. N. (1969). "Plant Pathology." Academic Press, New York.

Agrios, G. N. (1978). "Plant Pathology," 2nd ed. Academic Press, New York.

Akeley, R. V., Stevenson, F. J., and Schultz, E. S. (1948). Kennebec: A new potato variety resistant to late blight, mild mosaic, and net necrosis. *Am. Potato J.* **25**, 351–361.

Akeley, R. V., Stevenson, F. J., Schultz, E. S., Bonde, R., Nielsen, K. F., and Hawkins, A. (1955a). Saco: A new late-maturing variety of potato, immune from common race of the late blight fungus, highly resistant to if not immune from net necrosis and immune from mild and latent mosaics. *Am. Potato J.* **32**, 41–48.

Akeley, R. V., Stevenson, F. J., Blood, P. T., Schultz, E. S., Bonde, R., and Nielsen, K. F. (1955b). Merrimack: A new variety of potato resistant to late blight and ring rot adapted to New Hampshire. *Am. Potato J.* **32**, 93–99.

Akeley, R. V., Perry, B. A., and Schark, A. E. (1961). Redskin, a new red variety of potato resistant to scab, with high yielding ability, and adapted to growing conditions in the South. *Am. Potato J.* **38**, 81–84.

Akeley, R. V., Houghland, G. V. C., and Schark, A. E. (1962a). Genetic differences in potato-tuber greening. *Am. Potato J.* **39**, 409–417.

Akeley, R. V., Schark, A. E., McCubbin, E. N., and Eddins, A. H. (1962b). Ona, a new potato variety resistant to late blight, scab, Verticillium wilt and mild mosaic. *Am. Potato J.* **39**, 464–467.

Akeley, R. V., Perry, B., and Peel, R. D. (1968). Alamo: A new early- maturing, high yielding, widely adapted potato variety. *Am. Potato J.* **45**, 139–141.

Akeley, R. V., Murphy, H. J., and Cetas, R. C. (1971). Abnaki: A new high-yielding potato variety resistant to Verticillium wilt and leafroll. *Am. Potato J.* **48**, 230–233.

Alexopoulos, C. J. (1952). "Introductory Mycology." Wiley, New York.

Alfieri, S. A., and Stouffer, R. F. (1957). Evidence of immunity from virus S in the potato variety Saco. *Phytopathology* **47**, 1 (Abstr.).

Allen, J. D. (1957). The development of potato skin-spot disease. *Ann. Appl. Biol.* **45**, 293–298.

Alphey, T. J. W., Cooper, J. I., and Harrison, B. D. (1975). Systemic nematicides for the control of trichodorid nematodes and of potato spraing disease caused by tobacco rattle virus. *Plant Pathol.* **24**, 117–121.

Ambrosov, A. L., and Sokolova, L. A. (1976). Protiv virusnykh bolenznei kartofelya. [Against virus diseases of potato.] *Zashch. Rast. (Moscow)* **12**, 22–23.

Anonymous (1960). "Index of Plant Diseases in the United States," Agr. Handbook No. 165. U.S. Dept. of Agriculture, Washington, D.C.

Anonymous (1964). Monona, a new potato variety. *Am. Potato J.* **41**, 382.

Atanasoff, D. (1926). Sprain or internal brown spot of potatoes. *Phytopathology* **16**, 711–712.

Ayers, G. W. (1950). Fusarium storage rot of potatoes. *Proc. Can. Phytopathol. Soc.* **17**, 11–12 (Abstr.).

Ayers, G. W. (1961a). The susceptibility of potato varieties to storage rots caused by *Fusarium sambucinum* Fckl. F. 6 Wr. and *Fusarium caeruleum* (Lib.) Sacc. *Can. Plant Dis. Surv.* **41**, 170–171.

Ayers, G. W. (1961b). The susceptibility of potato varieties to wilt caused by *Verticillium alboatrum* Reinke and Berth. *Can. Plant Dis. Surv.* **41**, 172–173.

Ayers, G. W. (1974). Potato seed treatment for the control of Verticillium wilt and Fusarium seed piece decay. *Can. Plant Dis. Surv.* **54**, 74–76.

Baerecke, M. L. (1967). Prufung von Saco and Saco-Kreuzungen auf Resistenz gegen das S-Virus der Kartoffel. [Tests of Saco and Saco-crosses for resistance to potato virus S.] *Eur. Potato J.* **10**, 206–219.

Bagnall, R. H., and Larson, R. H. (1957). Potato virus S. *Phytopathology* **47**, 2–3.

Bagnall, R. H., and Young, D. A. (1959). Inheritance of immunity to virus S in the potato. *Am. Potato J.* **36**, 292.

Bagnall, R. H., Larson, R. H., and Walker, J. C. (1956). Potato viruses M, S, and X in relation to interveinal mosaic of the Irish Cobbler variety. *Res. Bull.—Wis., Agric. Exp. Stn.* **198**.

Bagnall, R. H., Wetter, C., and Larson, R. H. (1959). Differential host and serological relationships of potato virus M, potato virus S and carnation latent virus. *Phytopathology* **49**, 435–442.

Bannon, E. (1975). Susceptibility of potato cultivars to skin spot disease. *Potato Res.* **18**, 531–538.

Banville, G. J. (1978). Studies on the Rhizoctonia disease of potatoes. *Am. Potato J.* **55**, 56.

Barclay, G. M., Murphy, H. J., Manzer, F. E., and Hutchinson, F. E. (1973). Effects of differential rates of nitrogen and phosphorus on early blight in potatoes. *Am. Potato J.* **50**, 42–48.

Barrus, M. F. (1944). A Thecaphora smut on potatoes. *Phytopathology* **34**, 712–714.

Barrus, M. F., and Chupp, C. C. (1922). Yellow dwarf of potatoes. *Phytopathology* **12**, 123–132.

Barrus, M. F., and Muller, A. S. (1943). An Andean disease of potato tubers. *Phytopathology* **33**, 1086–1089.

Bawden, F. C. (1964). "Plant Viruses and Virus Diseases," 4th ed. Ronald Press, New York.

Bawden, F. C., and Sheffield, F. M. L. (1944). The relationship of some viruses causing necrotic diseases of the potato. *Ann. Appl. Biol.* **31**, 33–40.

Bazan De Segura, C. (1960). The gangrena disease of potato in Peru. *Plant Dis. Rep.* **44**, 257.

Bazan De Segura, C., and Del Carpio, R. (1974). La gangrena de la papa en la costa central peruana. [Potato gangrene on the central Peruvian coast.] *Informe Ministerio de Agricultura,* No. 35. *Rev. Plant Pathol.* **54**, 641.

Beaumont, A. (1947). Dependence on the weather of the dates of outbreak of potato blight epidemics. *Br. Mycol. Soc. Trans.* **31**, 45–53.

Beckman, C. H. (1973). The incidence of *Verticillium* species in soils, vines and tubers of Rhode Island-grown potatoes. *Plant Dis. Rep.* **57,** 928–932.

Beckman, C. H., Stessel, G. J., and Howard, F. L. (1969). *Verticillium* spp. and associated fungi from certified potato seed tubers. *Plant Dis. Rep.* **53,** 771–773.

Bennett, F. T. (1946). Soft rot of potatoes in 1945 crops. *Agriculture* **53,** 56–58.

Berger, K. C., and Gerloff, G. C. (1947). Stem streak necrosis of potatoes in relation to soil acidity. *Am. Potato J.* **24,** 156–162.

Biehn, W. L. (1970). Control of Verticillium wilt of potato by soil treatment with benomyl. *Plant Dis. Rep.* **54,** 171–173.

Bilbruck, J. D., and Rich, A. E. (1961). The effect of various dichlone treatments on the growth, yield, and disease incidence of potatoes and tomatoes in New Hampshire. *Plant Dis. Rep.* **45,** 128–133.

Black, L. M. (1937). A study of potato yellow dwarf in New York. *Cornell Univ. Agric. Exp. Stn. Memoir.* **209.**

Black, L. M. (1938). Properties of the potato yellow-dwarf virus. *Phytopathology* **28,** 863–874.

Black, L. M., and Price, W. C. (1940). The relationship between viruses of potato calico and alfalfa mosaic. *Phytopathology* **30,** 444–447.

Black, W. (1957). Incidence of physiological races of *Phytophthora infestans* in various countries. *Rep. Scot. Pl. Breed. Sta.,* 1957, pp. 43–49.

Black, W. (1960). Races of *Phytophthora infestans* and resistance problems in potatoes. *Rep. Scot. Plant Breed. Sta.,* 1960, pp. 29–38.

Black, W., and Malcolmson, J. F. (1965). New races of *Phytophthora infestans* (Mont.) de Bary and their complementary R-genes from *Solanum demissum* Lindl. *Am. Potato J.* **42,** 293–306.

Black, W., Masterbroek, C., Mills, W. R., and Peterson, L. C. (1953). A proposal for an international nomenclature of races of *Phytophthora infestans* and of genes controlling immunity in *Solanum demissum* derivatives. *Euphytica* **2,** 173–179.

Blodgett, E. C. (1943). Stem nematode on potato: A new potato disease in Idaho. *Plant Dis. Rep.* **27,** 658–659.

Blodgett, E. C., and Rich, A. E. (1950). Potato tuber diseases, defects and insect injuries in the Pacific Northwest. *Wash., Agric. Exp. Stn. Pop. Bull.* **195.**

Blodgett, F. M., and Stevenson, F. J. (1946). The new scab-resistant potatoes, Ontario, Seneca, and Cayuga. *Am. Potato J.* **23,** 315–329.

Bodleander, K. B. A., Lugt, C., and Marinus, J. (1964). The induction of second growth in potato tubers. *Eur. Potato J.* **7,** 57–71.

Boerema, G. H. (1967). The Phoma organisms causing gangrene of potatoes. *Neth. J. Plant Pathol.* **73,** 190–192.

Boesewinkel, H. J. (1976). Mattery-eye still threatens potato crops. *N. Z. J. Agric.* **132,** 21–22.

Bolkan, H. A. (1976). Seed tuber treatment for the control of black scurf disease of potatoes. *N. Z. J. Exp. Agric.* **4,** 357–361.

Bondarenko, E. E. (1973). Rezul'taty izucheniya stepeni ustoĭchivosti kartofelya k vozbuditelya raka *Synchytrium endobioticum* (Schilb.). Perc. [Results of studying the degree of resistance of potato to the causal agent of wart, *Synchytrium endobioticum* (Schilb.) Perc.] *Tr. Vses. Nauchno-Issled. Inst. Zashch. Rast.* **36,** 95–97; (*Rev. Plant Pathol.* **56,** 84).

Bonde, M. R., and McIntyre, G. A. (1968). Isolation and biology of a *Streptymyces* sp. causing potato scab in soils below pH 5.0. *Am. Potato J.* **45,** 273–278.

Bonde, R. (1939a). Bacterial wilt and soft rot on the potato in Maine. *Maine, Agric. Exp. Stn., Bull.* **396.**

Bonde, R. (1939b). Comparative studies of the bacteria associated with potato blackleg and seed-piece decay. *Phytopathology* **29,** 831–851.

Bonde, R. (1942). Ring rot in volunteer plants. *Am. Potato J.* **19**, 131-133.

Bonde, R. (1955). The effect of powdery scab on the resistance of potato tubers to late blight rot. *Maine, Agric. Exp. Stn., Bull.* **538.**

Bonde, R., and Hyland, F. (1960). Effects of antibiotic and fungicidal treatments on wound periderm formation, plant emergence, and yields produced by cut seed potatoes. *Am. Potato J.* **37**, 279-288.

Bonde, R., and Malcolmson, J. F. (1956). Studies in the treatment of potato seed pieces with antibiotic substances in relation to bacterial and fungous decay. *Plant Dis. Rep.* **40**, 615-619.

Bonde, R., and Merriam, D. (1951). Potato spindle tuber control. *Maine, Agric. Exp. Stn., Bull.* **487.**

Bonde, R., and Merriam, D. (1957). Knobby tuber disease of the potato. *Am. Potato J.* **34**, 227-229.

Bonde, R., and Schultz, E. S. (1943). Potato cull piles as a source of late blight infection. *Am. Potato J.* **20**, 112-118.

Bonde, R., and Schultz, E. S. (1945). The control of potato late blight tuber rot. *Am. Potato J.* **22**, 163-167.

Bonde, R., and Schultz, E. S. (1953). Purple top wilt and similar diseases of the potato. *Maine, Agric. Exp. Stn., Bull.* **511.**

Boyd, A. E. W. (1957). Field experiments on potato skin spot disease caused by *Oospora pustulans* Owen & Wakef. *Ann. Appl. Biol.* **45**, 284-292.

Boyd, A. E. W., and Lennard, J. H. (1961). Some effects of potato skin spot *Oospora pustulans* in Scotland. *Eur. Potato J.* **4**, 361-377.

Brandes, G. A., Bonde, R., Cetas, R. C., Samson, R. W., and Rich, A. E. (1959a). A review of uniform dosage and timing experiments comparing maneb and zineb fungicides on potatoes in Maine, New York, Indiana, and New Hampshire—1958. *Plant Dis. Rep.* **43**, 201-212.

Brandes, J., Wetter, C., Bagnall, R. H., and Larson, R. H. (1959b). Size and shape of the particles of potato virus S, potato virus M, and carnation latent virus. *Phytopathology* **49**, 443-446.

Brcak, J., Kralik, O., Limberk, J., and Ulrychova, M. (1969). Mycoplasmalike bodies in plants infected with potato witches' broom disease and the response of plants to tetracycline treatment. *Biol. Plant.* **11**, 470-476, *Rev. Plant Pathol.* **49**, 304.

Breed, R. S., Murray, E. G. D., and Smith, N. R., eds. (1957). "Bergey's Manual of Determinative Bacteriology," 7th ed. Williams & Wilkins, Baltimore, Maryland.

Brennan, E. I., Leone, I. A., and Daines R. H. (1964). The importance of variety in ozone plant damage. *Plant Dis. Rep.* **48**, 923-924.

Brown, E. B., and Sykes, G. B. (1973). Control of tobacco rattle virus (spraing) in potatoes. *Ann. Appl. Biol.* **75**, 462-464.

Burkholder, W. H., and Smith, W. L., Jr. (1949). *Erwinia atroseptica* (Van Hall) Jennison and *Erwinia carotovora* (Jones) Holland. *Phytopathology* **39**, 887-897.

Busch, L. V. (1965). Verticillium wilt of potato in Ontario. *Proc. Can. Phytopathol. Soc.* **32**, 10-19 (Abstr.).

Busch, L. V. (1966a). Effect of Di-Syston on control of Verticillium wilt of Kennebec potatoes. *Am. Potato J.* **43**, 286-288.

Busch, L. V. (1966b). Susceptibility of potato varieties to Ontario isolates of *Verticillium alboatrum. Am. Potato J.* **43**, 439-449.

Busch, L. V. (1973). Effects of some cropping practices on survival of *Verticillium. Am. Potato J.* **50**, 381-382.

Busch, L. V., and Ashton, G. C. (1964). Chemical control of common scab of potato. *Am. Potato J.* **41**, 92-94.

Callbeck, L. C. (1960). Screening of potato fungicides in 1960. *Canad. Pl. Dis. Surv.* **40,** 56–58.

Cammack, R. H. (1964). The abnormality resembling potato spindle tuber in the variety Redskin. *Plant Pathol.* **13,** 69–72.

Campbell, W. P., and Griffiths, D. A. (1973). Pathogenicity of *Verticillium dahliae* to potato in Victoria, Australia. *Plant Dis. Rep.* **57,** 735–738.

Campbell, J. C., and Young, D. A. (1970). Raritan, a potato with high specific gravity. *Am. Potato J.* **47,** 264–267.

Cannon, F. M., and Callbeck, L. C. (1965). High versus low volume spraying of potatoes. *Am. Potato J.* **42,** 328–332.

Catchpole, A. H., and Hillman, J. R. (1975a). Studies of the coiled sprout disorder of the potato. 1. Description of the disorder. *Potato Res.* **18,** 282–289.

Catchpole, A. H., and Hillman, J. R. (1975b). Studies of the coiled sprout disorder of the potato. 2. Effects of sprout length and inoculation with *Verticillium nubilum* under field conditions. *Potato Res.* **18,** 539–545.

Catchpole, A. H., and Hillman, J. R. (1975c). Studies of the coiled sprout disorder of the potato. 3. Effects of sprout length, planting depth, and growth media. *Potato Res.* **18,** 596–607.

Cervantes, J. (1965). Late blight resistance of nine Mexican Potato varieties in ten years of field trials. *Am. Potato J.* **42,** 258 (Abstr.).

Cetas, R. C. (1970). Interaction of seed piece treatments and soil treatments of aldicarb and benomyl in the control of Verticillium-nematode complex of potatoes. *Phytopathology* **60,** 572 (Abstr.).

Cetas, R. C. (1971). Lack of correlation between pink eye and Verticillium wilt susceptibility among potato cultivars and breeding lines. *Am. Potato J.* **48,** 306 (Abstr.).

Cetas, R. C., and Sawyer, R. L. (1962). Evaluation of Uracide for the control of common scab of potatoes on Long Island. *Am. Potato J.* **39,** 456–459.

Chitwood, B. G. (1951). The golden nematode of potatoes. *U.S., Dep. Agric., Circ.* **875.**

Chupp, C. (1953). "A Monograph of the Fungus Genus *Cercospora.*" Published by the Author, Ithaca, New York.

Chupp, C., and Sherf, A. F. (1960). "Vegetable Diseases and Their Control." Ronald Press, New York.

Clarke, R. G. (1981). Potato leafroll virus purification and antiserum preparation for enzyme-linked immunosorbent assays. *Am. Potato J.* **58,** 291–298.

Clinch, P., Loughnane, J. B., and Murphy, P. A. (1936). A study of the aucuba or yellow mosaics of the potato. *Sci. Proc. R. Dublin Soc.* **21,** 431–438.

Cole, H., Mills, W. R., and Massie, L. B. (1972). Influence of chemical seed and soil treatment on Verticillium-induced yield reduction and tuber defects. *Am. Potato J.* **49,** 79–92.

Conners, I. L. (1967). "An Annotated Index of Plant Diseases in Canada." *Canada Dept. of Agr. Res. Branch Publ.* 1251.

Conroy, R. J. (1954). Purple-top wilt of potato. *N. S. W., Dep. Agric., Sci., Bull.* **75.**

Cooper, J. I., and Thomas, P. R. (1971). Chemical treatment of soil to prevent transmission of tobacco rattle virus in potatoes by *Trichodorus* spp. *Ann. Appl. Biol.* **69,** 23–34.

Copeland, R. B., and Logan, C. (1975). Control of tuber diseases, especially gangrene, with benomyl, thiabendazole and other fungicides. *Potato Res.* **18,** 179–188.

Cox, H. E., and Large, E. C. (1960). Potato blight epidemics throughout the world, Agr. Handbook No. 174. U.S. Dept. of Agriculture, Washington, D.C.

Cox, R. S. (1965). The role of potato-Y virus in tomato production in South Florida. *Plant Dis. Rep.* **49,** 1018.

Cromarty, R. W., and Easton, G. D. (1973). The incidence of decay and factors affecting bacterial soft rot of potatoes. *Am. Potato J.* **50,** 398–407.

Crumbly, I. J., Nelson, D. C., and Duysen, M. E. (1973). Relationship of hollow heart in Irish potatoes to carbohydrate reabsorption and growth rate of tubers. *Am. Potato J.* **50**, 266–274.

Cunningham, C. E., Akeley, R. V., Peterson, L. C., and Snyder, T. E. (1968). Wauseon: A new potato variety resistant to golden nematode with good processing quality. *Am. Potato J.* **45**, 146–149.

Cunningham, H. S. (1936). The root-knot nematode *(Heterodera marioni)* in relation to the potato industry on Long Island. *N.Y., Agric. Exp. Stn., Geneva,* Bull. **667**.

Cunningham, H. S., and Mai, W. F. (1947). Nematodes parasitic on the Irish potato. *Cornell Exten. Bull.* **712**.

Cunningham, H. S., and Reinking, O. A. (1946). Fusarium seed piece decay of potato on Long Island and its control. *N.Y., Agric. Exp. Stn., Geneva,* Bull. **721**.

Dale, E. (1912). On the cause of blindness in potato tubers. *Ann. Bot. (London)* **26**, 129–131.

Dallimore, C. E. (1972). Control of corky ringspot in Russet Burbank potatoes by soil fumigation. *Am. Potato J.* **49**, 366 (Abstr.).

Darby, J. F., Larson, R. H., and Walker, J. C. (1951). Variation in virulence and properties of potato virus Y strains. *Wis. Agric. Exp. Stn. Res. Bull.* **177**.

Darling, H. M. (1957). Control of the potato rot nematode in Wisconsin. *Phytopathology* **47**, 7 (Abstr.).

Darling, H. M. (1959). North American potato varieties. *Potato Assn. of America Handb.* **4**, 19–41.

Darling, H. M. (1977). Seed potato certification. *In* "Potatoes: Production, Storing, Processing," 2nd ed. by O. Smith, pp. 405–416. Avi Publ., Westport, Connecticut.

Davidson, R. S. (1946). Ring-rot-like symptoms produced by soft-rot bacteria in potato tubers. *Phytopathology* **36**, 237 (Abstr.).

Davidson, R. S. (1948). Factors affecting the development of bacterial soft rot of potato tuber initials. *Phytopathology* **38**, 673–687.

Davidson, T. R., and Sanford, G. B. (1955). Expression of leafroll phloem necrosis in potato tubers. *Can. J. Agric. Sci.* **35**, 42–47.

Davies, H. T., McEwen, H. L., and Dixon, N. C. (1975). Field testing potatoes for resistance to leafroll and virus Y. *Am. Potato J.* **52**, 151–155.

Davies, H. T., and Young, L. C. (1966). Sable, a new early variety of potato. *Am. Potato J.* **43**, 154–157.

Davies, H. T., Young, D. A., Munro, J., and Young, L. C. (1963). Hunter, a new potato variety with excellent cooking quality and field immune to viruses X and A. *Am. Potato J.* **40**, 275–278.

Davis, J. R. (1973). Seed and soil treatments for control of *Rhizoctonia* and blackleg of potato. *Plant Dis. Rep.* **57**, 803–806.

Davis, J. R. (1978). The Rhizoctonia disease of potato in Idaho. *Am. Potato J.* **55**, 58–59.

Davis, J. R., and Allen, T. C. (1975). Weed hosts of the tobacco rattle virus in Idaho. *Am. Potato J.* **52**, 1–8.

Davis, J. R., McMaster, G. M., Garner, J. G., and Callihan, R. H. (1972). Effects of soil moisture and fungicide treatments on potato scab. *Am. Potato J.* **49**, 360 (Abstr.).

Davis, J. R., Garner, J. G., and Callihan, R. H. (1974). Effects of gypsum, sulfur, terraclor and terraclor super-X for potato scab control. *Am. Potato J.* **51**, 35–43.

Dearborn, C. H. (1964). Alaska Russet: A new potato for Alaska. *Am. Potato J.* **41**, 137–139.

Dearborn, C. H. (1969). Alaska Frostless, an inherent frost resistant potato variety. *Am. Potato J.* **46**, 1–4.

DeBokx, J. A. (1969). Particle length of various isolates of potato virus S. *Neth. J. Plant Pathol.* **75**, 144–146.

Dewey, D. H., and Barger, W. R. (1948). The occurrence of bacterial soft rot on potatoes resulting from washing in deep vats. *Proc. Am. Soc. Hortic. Sci.* **52**, 325–330.

Dickerson, O. J., Darling, H. M., and Griffin, G. D. (1964). Pathogenicity and population trends of *Pratylenchus penetrans* on potato and corn. *Phytopathology* **54**, 317–322.

Dickson, B. T. (1926). The "black dot" disease of potato. *Phytopathology* **16**, 23–24.

Diener, T. O. (1975). State of the viroid: RNA characterization. *Am. Potato J.* **52**, 242 (Abstr.).

Dinkel, D. H. (1960). A study of factors influencing the development of hollow heart in Irish Cobbler potatoes (*Solanum tuberosum* L.). *Diss. Abstr.* **21**, 14–15.

Doi, Y., Teravaka, M., Yora, K., and Asuyama, H. (1967). Mycoplasma, or PLT group-like micro-organisms, found in the phloem elements of plants infected with mulberry dwarf, potato witch's broom, aster yellows, or Paulownia witch's broom. *Ann. Phytopathol. Soc. Jpn.* **33**, 259–266.

Douglas, D. R., and Groskopp, M. D. (1974). Control of early blight in eastern and south central Idaho. *Am. Potato J.* **51**, 361–368.

Dowson, W. J. (1957). "Plant Diseases Due to Bacteria," 2nd ed. Cambridge Univ. Press, London and New York.

Drechsler, C. (1919). Morphology of the genus *Actinomyces. Bot. Gaz. (Chicago)* **67**, 65–83; 147–168.

Duffus, J. E. (1981a). Beet western yellows virus—a major component of some potato leaf roll-affected plants. *Phytopathology* **71**, 193–196.

Duffus, J. E. (1981b). Distribution of beet western yellows virus in potatoes affected by potato leaf roll. *Plant Disease* **65**, 819–820.

Duffus, J. E., and Gold, A. H. (1969). Membrane feeding and infectivity neutralization used in a serological comparison of potato leaf roll and beet western yellows viruses. *Virology* **37**, 150–153.

Duncan, J., and Généreux, H. (1960). La Transmission par les insectes de *Corynebacterium sepedonicum* (Spieck. & Kott.) Skaptason et Burkholder. [Transmission of *C. sepedonicum* by insects.] *Can. J. Plant Sci.* **40**, 110–116.

Dutt, B. L., Rai, R. P., and Harikoshore, H. (1973). Evaluation of reaction of potato to powdery-mildew. *Indian J. Agric. Sci.* **43**, 1063–1066.

Dwelle, R. B., Stallknecht, G. F., McDole, R. E., and Pavek, J. J. (1975). Effects of soil potash treatment and storage temperature on blackspot bruise susceptibility of four *Solanum tuberosum* cultivars. *Am. Potato J.* **52**, 277 (Abstr.).

Dykstra, T.P. (1948). Potato diseases and their control. *U.S. Dept. Agr. Farmers' Bull.* **1881**.

Dykstra, T. P., and Reid, W. J., Jr. (1956). Potato growing in the South. *U.S. Dept. Agr. Farmers' Bull.* **2098**.

Dykstra, T. P., Gilmore, T. R., and Miller, J. C. (1961). Catoosa: A new medium-early red potato immune from common races of the late blight fungus and resistant to scab. *Am. Potato J.* **38**, 300–303.

Eastman, P. J. (1976). Seed information. *Potato Counc.* **22**(10), 3, 5.

Easton, G. D. (1978). The Rhizoctonia disease of potato in Washington. *Am. Potato J.* **55**, 57–58.

Easton, G. D., Larson, R. H., and Hougas, R. W. (1958). Immunity to virus X in the genus *Solanum Res. Bull.—Wis., Agric. Exp. Stn.,* **205.**

Easton, G. D., and Nagle, M. E. (1981). Potato late blight control by the systemic fungicide Ridomil. *Phytopathology* **71**, 214 (Abstr.).

Easton, G. D., Nagle, M. E., and Bailey, D. L. (1969). Effect of repeated annual field fumigations and vine-burnings on *Verticillium* in infested soil. *Phytopathology* **59**, 1025 (Abstr.).

Easton, G. D., Nagle, M. E., and Bailey, D. L. (1970). Potato seed piece treatment in Washington. *Am. Potato J.* **37**, 469–474.

Easton, G. D., Nagle, M. E., and Bailey, D. L. (1972). *Verticillium albo-atrum* carried by certified seed potatoes into Washington and control of this pathogen by chemicals. *Am. Potato J.* **49**, 397–402.

Easton, G. D., Nagle, M. E., and Bailey, D. L. (1974). Fumigants, rates, and application methods affecting Verticillium wilt incidence and potato yields. *Am. Potato J.* **51**, 71–77.

Easton, G. D., Nagle, M. E., and Bailey, D. L. (1975). Lack of foliar protection from early blight by aircraft-applied fungicides on sprinkler-irrigated potatoes. *Plant Dis. Rep.* **59**, 910–914.

Eddins, A. H. (1936). Brown rot of Irish potatoes and its control. *Bull.—Fla., Agric. Exp. Stn.* **299**.

Eddins, A. H. (1939). Adjusting pH reaction in soils with sulphur and limestone to control brown rot of potatoes. *Am. Potato J.* **16**, 6–16.

Eddins, A. H., Proctor, E. Q., and West, E. (1946). Corky ring spot of potatoes in Florida. *Am. Potato J.* **23**, 330–333.

Edgington, L. V. (1962). Influence of Connecticut temperatures on the relation of pathogenicity of Maine and Connecticut Verticillium isolates. *Am. Potato J.* **39**, 261–265.

Edgington, L. V., and Busch, L. V. (1967). Control of Rhizoctonia stem canker in potato. *Can. Plant Dis. Surv.* **47**, 28–29.

Edmundson, W. C., McLean, J. G., Frutchey, C. W., and Schaal, L. A. (1961). Navajo and Blanca: Two new potato varieties resistant to scab and adapted to Colorado. *Am. Potato J.* **38**, 236–239.

Edwardson, J. R. (1974a). Some properties of the potato virus Y-group. *Monogr. Ser.—Fla., Agric. Exp. Stn.* **4**, 1–398.

Edwardson, J. R. (1974b). Host ranges of viruses in the PVY-group. *Monogr. Ser.—Fla., Agric. Exp. Stn.* **5**, 225.

Elliott, C. (1951). "Manual of Bacterial Plant Pathogens," 2nd ed. Chronica Botanica, Waltham, Massachusetts.

Ellison, J. H., and Jacob, W. C. (1952). Internal browning of potatoes as affected by date of planting and storage. *Am. Potato J.* **29**, 241–252.

Engelhard, A. W. (1957) Host index of *Verticillium albo-atrum* Heinke & Berth. (including *Verticillium dahliae* Kleb.). *Plant Dis. Rep. Suppl.* 244.

Erickson, H. T. (1960). Potato scab control on organic soils. I. Initial responses to PCNB. *Am. Potato J.* **37**, 18–22.

Erinle, I. D. (1975a). Blackleg of potatoes: Induction through tuber inoculation. *Plant Pathol.* **24**, 172–175.

Erinle, I. D. (1975b). Growth of *Erwinia carotovora* var. *atroseptica* and *E. carotovora* var. *carotovora* in potato stems. *Plant Pathol.* **24**, 224–229.

Evans, K., and Brodie, B. B. (1980). The origin and distribution of the golden nematode and its potential in the U.S.A. *Am. Potato J.* **57**, 79–89.

Feddersen, H. D. (1962). Target spot of potatoes. Trials show value of spraying. *J. Agric. South Aust.* **65**, 300–308.

Fernow, K. H., Peterson, L. C., and Plaisted, R. L. (1970). Spindle tuber virus in seeds and pollen of infected potato plants. *Am. Potato J.* **47**, 75–80.

Fink, H. C. (1957). Potato late blight control in Pennsylvania, 1956, *Phytopathology,* **47**, 241–242.

Foister, C. E. (1952). The distribution and prevalence of potato gangrene. *Plant Pathol.* **1**, 85.

Folsom, D. (1946). Potato yellowtop and unmottled curly dwarf in Maine. *Maine, Agric. Exp. Stn. Bull.* **446**.

Folsom, D. (1947). Permanence of greening of potato tubers. *Am. Potato J.* **24**, 336–340.

Folsom, D. (1953). Verticillium wilt in potatoes. *Maine Farm Res.* **1**,(2), 9–11.

Folsom, D., and Bonde, R. (1925). *Alternaria solani* as a cause of tuber rot in potatoes. *Phytopathology* **15**, 282–286.

Folsom, D., and Bonde, R. (1936). List of distinct potato viruses. *Am. Potato J.* **13**, 14–16.

Folsom, D., and Friedman, B. A. (1959). *Pseudomonas fluorescens* in relation to certain diseases of potato tubers in Maine. *Am. Potato J.* **36**, 90–97.

Folsom, D., and Rich, A. E. (1940). Potato tuber net necrosis and stem-end browning studies in Maine. *Phytopathology* **30**, 313–322.

Folsom, D., and Stevenson, F. J. (1946). Resistance of potato seedling varieties to the natural spread of leaf roll. *Am. Potato J.* **23**, 247–264.

Folsom, D., Getchell, J. S., and Bonde, R. (1948). Bacterial red xylem disease of potato tubers in Maine. *Plant Dis. Rep.* **32**, 230–231.

Folsom, D., Wyman, O. L., and E. Westin. (1951). Potato Verticillium wilt. *Univ. of Maine Ext. Circ.* 265.

Folsom, D., Simpson, G. W., and Bonde, R. (1955). Maine potato diseases, insects, and injuries. *Maine, Agric. Exp. Stn., Bull.* **469.**

Francki, R. I. B. (1967). Effect of high light intensities on spontaneous and virus-induced local lesions in *Gomphrena globosa. Phytopathology* **57**, 329 (Abstr.).

Frank, J. A. (1975). The relative importance of potato tuber-borne Rhizoctonia inoculum in comparison to soil-borne inoculum. *Am. Potato J.* **52**, 244 (Abstr.).

Frank, J. A. (1978). The Rhizoctonia disease of potatoes in Maine. *Am. Potato J.* **55**, 59–60.

Frank, J. A., and Francis, S. K. (1976). The effect of a *Rhizoctonia solani* phytotoxin on potatoes. *Can. J. Bot.* **54**, 2536–2540.

Frank, J. A., and Murphy, H. J. (1977). The effect of crop rotations on Rhizoctonia disease of potatoes. *Am. Potato J.* **54**, 315–322.

Frank, J. A., Webb, R. E., and Wilson, D. R. (1973). The relationship between Verticillium wilt and the pinkeye disease of potatoes. *Am. Potato J.* **50**, 431–438.

Frank, J. A., Leach, S. S., and Webb, R. E. (1976). Evaluation of potato clone reaction to *Rhizoctonia solani. Plant Dis. Rep.* **60**, 910–912.

Fribourg, C. E. and de Zoeten, G. A. (1975). Some leguminous hosts for potato virus X. *Plant Dis. Rep.* **59**, 923–925.

Fribourg, C. E., Jones, R. A. C., and Koenig, R. (1977). Host plant reactions, physical properties and serology of three isolates of Andean potato latent virus from Peru. *Ann. Appl. Biol.* **86**, 373–380.

Friedman, B. A., and Folsom, D. (1953). Storage behavior of Kennebec potatoes infected by *Verticillium albo-atrum. Phytopathology* **43**, 108.

Fry, W. E. (1977). Integrated control of potato late blight: Effects of polygenic resistance and techniques of timing fungicide applications. *Phytopathology* **67**, 415–420.

Galindo, J. A. (1965a). Sexuality and genetics of *Phytophthora infestans. Am. Potato J.* **42**, 256–265.

Galindo, J. A. (1965b). Sexuality and genetics of *Phytophthora infestans. Am. Potato J.* **42**, 293–306.

Gallegly, M. E., and Galindo, J. (1957). The sexual stage of *Phytophthora infestans* in Mexico. *Phytopathology* **47**, 13 (Abstr.).

Gerrard, E. H. (1946). A storage rot of potatoes caused by a fluorescent organism resembling *Pseudomonas fluorescens* (Flugge) Migula. *Can. J. Res., Sect. C* **24**, 79–84.

Gibbs, A. J., and Harrison, B. D. (1969). Eggplant mosaic virus, and its relationship to Andean potato latent virus. *Ann. Appl. Biol.* **64**, 225–231.

Gibbs, A. J., Hecht-Poinar, J. E., Woods, R. D., and McKee, R. K. (1966). Andean potato latent, Dulcamara mottle and Ononis yellow mosaic. *J. Gen. Microbiol.* **44**, 177–193.

Gibson, R. W. (1974). The induction of top-roll symptoms on potato plants by the aphid *Macrosiphum euphorbiae. Ann. Appl. biol.* **76**, 19–26.

Gibson, R. W. (1975). Potato seed tubers do not transmit top-roll. *Plant Pathol.* **24**, 107–108.

Gilbert, A. H. (1928). Net necrosis of Irish potato tubers. *Vt., Agric. Exp. Stn. Bull.* **289**.

Goss, R. W. (1930). The symptoms of spindle tuber and unmottled curly dwarf of the potato. *Res. Bull.—Neb., Agric. Exp. Stn.* **47**.

Goss, R. W. (1931). Infection experiments with spindle tuber and unmottled curly dwarf on the potato. *Res. Bull.—Neb., Agric. Exp. Stn.* **53**.

Goss, R. W. (1949). Pink rot of potatoes caused by *Phytophthora erythroseptica* Pethyb. *Res. Bull.—Neb., Agric. Exp. Stn.* **160**.

Goss, R. W., and Jensen, J. H. (1942). Varietal susceptibility of potatoes to Fusarium wilt. *Am. Potato J.* **18**, 209–212.

Goth, R. W., and Webb, R. E. (1975). Lack of potato virus S transmission via true seed in *Solanum tuberosum. Phytopathology* **65**, 1347–1349.

Graham, D. C. (1962). Black leg disease of potatoes. *Rev. Appl. Mycol.* **42**, 46 (Abstr.).

Graham, D. C. (1964). Taxonomy of the soft rot coliform bacteria. *Annu. Rev. Phytopathol.* **2**, 13–42.

Graham, D. C., and Harper, P. C. (1966). Effect of inorganic fertilizers on the incidence of potato blackleg disease. *Eur. Potato J.* **9**, 141–145.

Graham, D. C., and Harrison, M. D. (1975). Potential spread of *Erwinia* spp. in aerosols. *Phytopathology* **65**, 739–741.

Gries, G. A., and Horsfall, J. G. (1946). Polymodal dosage-response curve between calcium-potassium ratio and potato scab. *Phytopathology* **36**, 399.

Gronquist, J. A., and Anderson, N. A. (1977). A glasshouse test to evaluate the response of potato cultivars to *Rhizoctonia solani. Abstr. 69th Annu. Am. Phytopathol. Soc.* 196 (Abstr.).

Guthrie, J. W. (1959a). Net necrosis of Russet Burbank tubers from seed-borne potato leaf-roll infected plants. *Phytopathology* **49**, 113.

Guthrie, J. W. (1959b). The early, dwarf symptoms of bacterial ring rot of potato in Idaho. *Phytopathology* **49**, 453–454.

Guthrie, J. W. (1961). Net necrosis associated with primary and secondary infection of leafroll virus in Russet Burbank. *Am. Potato J.* **38**, 435–439.

Hansen, A. A. (1925). Two fatal cases of potato poisoning. *Science* **61**, 340–341.

Harrison, B. D. (1964). The transmission of plant viruses in soil. *In* "Plant Virology" (M. K. Corbett and H. D. Sisler, eds.), pp. 118–147. Univ. of Florida Press, Gainesville.

Harrison, B. D. (1971). Potato viruses in Britain. *In* "Diseases of Crop Plants" (J. H. Western, ed.), pp. 123–159. Wiley, New York.

Harrison, B. D., and Roberts, I. M. (1969). Association of mycoplasmalike bodies with potato witch's broom disease from Scotland. *Ann. Appl. Biol.* **63**, 347–349.

Harrison, M. D. (1962). Potato russet scab, its cause and factors affecting its development. *Am. Potato J.* **39**, 368–387.

Harrison, M. D., and Venette, J. R. (1970). Chemical control of potato early blight and its effect on potato yield. *Am. Potato J.* **47**, 81–86.

Harrison, M. D., Livingston, C. H., and Oshima, N. (1965a). Epidemiology of potato early blight in Colorado. I. Initial infection, disease development, and the influence of environmental factors. *Am. Potato J.* **42**, 279–291.

Harrison, M. D., Livingston, C. H., and Oshima, N. (1965b). Control of potato early blight in Colorado. I. Fungicidal spray schedules in relation to the epidemiology of the disease. *Am. Potato J.* **42**, 319–327.

Harrison, M. D., Livingston, C. H., and Oshima, N. (1965c). Control of early blight in Colorado. II. Spore traps as a guide for initiating applications of fungicides. *Am. Potato J.* **42**, 333–340.

Haware, M. P. (1968). Assessment of losses due to early blight (*Alternaria solani*) on potato. *JNKVV Res. J.* **2**, 67–68.

Hawkes, J. G. (1978). History of the potato. *In* "The Potato Crop" (P. M. Harris, ed.), pp. 1–14. Chapman & Hall, London.

Hawkins, A., and Miller, P. M. (1971a). Row treatment of potatoes with systemics for meadow nematode (*Pratylenchus penetrans*) control. *Am. Potato J.* **48**, 21–25.

Hawkins, A., and Miller, P. M. (1971b). Row fumigation with Vorlex for control of a meadow nematode (*Pratylenchus penetrans*) in potatoes. *Am. Potato J.* **48**, 64–68.

Haynes, F. L., Stevenson, F. J., Akeley, R. V., and Cochran, F. D. (1956). Boone: A new variety of potato resistant to late blight and adapted to western North Carolina. *Am. Potato J.* **33**, 315–318.

Heggestad, H. E. (1970). Variations in response of potato cultivars to air pollution. *Phytopathology* **60**, 1015 (Abstr.).

Hellmers, E. (1959). *Pectobacterium carotovorum* var. *atrosepticum* (van Hall) Dowson the correct name of the potato black leg pathogen; a historical and critical review. *Eur. Potato J.* **2**, 251–271.

Henning, R. G., and Alexander, L. J. (1959). Evidence of existence of physiological races of *Alternaria solani*. *Plant Dis. Rp.* **43**, 298–308.

Henninger, M. R. (1979). Tuber necrosis in "Atlantic," *Am. Potato J.* **56**, 464 (Abstr.).

Hesen, J. C., and Kroesbergen, E. (1960). Mechanical damage to potatoes. I. *Eur. Potato J.* **3**, 30–46.

Hueberger, J. W., Davis, S. H., Jr., Nicholas, L. P., and Buehler, S. D. (1947). Zinc ethylene bisdithiocarbamate as a fungicide on vegetables. *Phytopathology* **37**, 9 (Abstr.).

Hide, G. A., and Corbett, D. C. M. (1973). Controlling early death of potatoes caused by *Heterodera rostochiensis* and *Verticillium dahliae*. *Ann. Appl. Biol.* **75**, 461–462.

Hide, G. A., Hirst, J. M., and Mundy, E. J. (1969). The phenology of skin spot (*Oospora pustulans* Owen & Wakef.) and other fungal diseases of potato tubers. *Ann. Appl. Biol.* **64**, 265–279.

Hines, M. (1976). The weather relationships of powdery scab disease of potatoes. *Ann. Appl. Biol.* **84**, 274–275.

Hirst, J. M., and Stedman, O. J. (1960). The epidemiology of *Phythophthora infestans*. I. Climate, ecoclimate, and the phenology of disease outbreak. *Ann. Appl. Biol.* **48**, 471–478.

Hodges, L. R. (1973). "Nematodes and Their Control." Union Carbide Corp., Salinas, California.

Hodgson, W. A., Pond, D. D., and Munro, J. (1974). Diseases and pests of potatoes. *Publ.—Can. Dep. Agric.* **1492**.

Holbrook, E., and Kleinschmidt, D. (1975). Potato pest management in Maine. *Am. Potato J.* **52**, 278 (Abstr.).

Hollis, J. P. (1949). Location of bacteria in healthy potato tissue. *Phytopathology* **39**, 9–10.

Holmes, F. O. (1948). Order Virales—The filterable viruses. *In* "Bergey's Manual of Determinative Bacteriology" (R. S. Breed, E. G. D. Murray, and A. P. Hitchens, eds.), 6th ed., Suppl. II, pp. 1125–1286. Williams & Wilkins, Baltimore, Maryland.

Hooker, W. J. (1957). "Control of Common Scab of Potatoes," Potato Handb. No. 2. Potato Assoc. Am., New Brunswick, New Jersey.

Hooker, W. J. (1973). Unusual aspects of lightning injury in potato. *Am. Potato J.* **50**, 258–265.

Hooker, W. J., ed. (1981). "Compendium of Potato Diseases." American Phytopathological Society, St. Paul, Minnesota.

Hooker, W. J., Yang, T. C., and Potter, H. S. (1972). Air pollution effects on potato and bean in southern Michigan. *Mich., Agric. Exp. Stn., Res. Rep.* **167.**

Hooker, W. J., Yang, T. C., and Potter, H. S. (1973). Air pollution injury of potato in Michigan. *Am. Potato J.* **50,** 151–161.

Houghland, G. V. C., and Cash, L. C. (1957). Carry-over effects of PCNB applied to the soil for the control of potato scab. *Am. Potato J.* **34,** 85–88.

Houghland, G. V. C. Akeley, R. V., Dykstra, T. P., and Shands, W. A. (1954). Potato production in the northeastern and north central states. *U.S. Dept. Agr. Farmers' Bull.* **1958.**

Howard, F. D., Laborde, J. F. Yamaguchi, M., and Knott, J. E. (1961). Studies of internal black spot of California-grown White Rose potato tubers. *Proc. Am. Soc. Hortic. Sci.* **78,** 406–412.

Hoyman, W. G. (1964). Red Pontiac vine and tuber yields as affected by virus X. *Am. Potato J.* **41,** 208–211.

Hoyman, W. G. (1970). Cascade: A new variety for the processing of frozen french fries. *Am. Potato J.* **47,** 261–263.

Hoyman, W. G., and Dingman, E. (1965). Effect of certain systemic insecticides on the incidence of Verticillium wilt and the yield of Russet Burbank potato. *Am. Potato J.* **42,** 195–200.

Hoyman, W. G., and Dingman, E. (1967). Temik: A systemic insecticide effective in delaying Verticillium wilt of potato. *Am. Potato J.* **44,** 3–8.

Hoyman, W. G., and Holland, R. C. (1974). Nooksack: A russet potato adapted to northwestern Washington. *Am. Potato J.* **51,** 99–102.

Hoyman, W. G., Picha, B., and Turnquist, O. C. (1963). Reliance: A new potato variety with moderate resistance to some common pathogens. *Am. Potato J.* **40,** 406–410.

Hudson, D. E., and Orr, P. H. (1977). Incidence of mechanical injury to potatoes during certain storage-related handling operations in the Red River Valley production area. *Am. Potato J.* **54,** 11–21.

Huguelet, J. E. (1976). Potato diseases and control. *In* "Potato Production in North Dakota," pp. 21–49. North Dakota State University Ext. Bull. 26.

Huisman, O. C., and Ashworth, L. J. Jr. (1976). Rotation ineffective as *Verticillium* control. *Calif. Agr.* **30** (11), 14–15.

Hunter, J. E. (1965). Studies on potato spindle tuber virus. *Diss. Abstr.* **25,** 4344–4345.

Hunter, J. E., and Rich, A. E. (1964). The effect of spindle tuber virus on growth and yield of Saco potatoes. *Am. Potato J.* **41,** 113–116.

Hussain, C. A. (1965). Effect of planting date and variety on yield and internal brown spot in potatoes in Lebanon. *West Park. J. Agric. Res.* **3,** 76–80.

Hyre, R. A. (1955). Three methods of forecasting late blight of potato and tomato in the northeastern United States. *Am. Potato J.* **32,** 362–371.

Hyre, R. A., and Bonde, R. (1955). Forecasting late blight of potato in northern Maine. *Am. Potato J.* **32,** 119–125.

Hyre, R. A., Bonde, R., and F. E. Manzer. (1960). Reevaluation in Maine of three methods proposed for forecasting late blight. *Plant Dis. Rep.* **44,** 235–237.

Idaho Crop Improvement Assn. (1978). 1978 Idaho Certified Seed Potatoes. Idaho Crop Improvement Assn., Boise, Idaho.

International Potato Center (1977). "The Potato: Major Diseases and Nematodes." Centro Internacional de la Papa, Lima, Perú.

Iritani, W. M., and Weller, L. (1973). The develoment of translucent end tubers. *Am. Potato J.* **50,** 223–233.

Ishihara, T. (1969). Families and genera of leafhopper vectors. *In* "Viruses, Vectors, and Vegetation" (K. Maramorosch, ed.), pp. 235–254. Wiley-Interscience, New York.

Jackson, A. W., and Henry, A. W. (1946). Occurrence of *Bacillus polymyxa* (Praz.) Mig. in Alberta soils with special reference to its pathogenicity on potato tubers. *Can. J. Res., Sect. C* **24,** 39–46.

Jacob, W. C. (1959). Studies on internal black spot of potatoes. *Cornell Univ. Agric. Exp. Stn. Mem.* **368.**

James, W. C., Bradley, R. H. E., Smith, C. S., and Wong, S. I. (1975). Misses in potato crops in New Brunswick in 1973; their extent, distribution and cause. *Am. Potato J.* **52,** 83–87.

Johansen, R. H. (1965). Norgold Russet, a new early maturing potato variety with good type and scab resistance. *Am. Potato J.* **42,** 201–204.

Johansen, R. H., Lana, E. P., and Benson, A. P. (1963). Snowflake, a new early-maturing white-skinned potato with field resistance to virus Y. *Am. Potato J.* **40,** 271–274.

Johansen, R. H., Schulz, J. T., and Huguelet, J. E. (1969a). Norchip, a new early maturing chipping variety with high total solids. *Am. Potato J.* **46,** 254–258.

Johansen, R. H., Schulz, J. T., and Huguelet, J. E. (1969b). Norchief, a new smooth type, high total solids, red-skinned potato variety. *Am. Potato J.* **46,** 298–301.

Johansen, R. H., Farnsworth, B., Huguelet, J. E., Nelson, D. C., and Lana, E. P. (1977). Bison, a new red-skinned potato variety. *Am. Potato J.* **54,** 189–193.

Johnston, G. R., Rowberry, R. G., and Mainprize, L. F. (1970). York, an early potato high in potato solids and adapted to organic soils. *Am. Potato J.* **47,** 201–204.

Jones, E. D., and Mullen, J. M. (1974). The effect of potato virus X on susceptibility of potato tubers to *Fusarium roseum 'Avenaceum.'* *Am. Potato J.* **51,** 209–215.

Jones, L. R. (1905). Disease resistance of potatoes. *U.S., Dep. Agric., Bur. Plant Ind. Bull.* **87.**

Jones, R. A. C., and Fribourg, C. E. (1977). Beetle, contact and potato true seed transmission of Andean potato latent virus. *Ann. Appl. Biol.* **86,** 123–128.

Juo, P., and Rich, A. E. (1965). Indicator hosts for potato aucuba mosaic. *Phytopathology* **55,** 1284 (Abstr.).

Juo, P., and Rich, A. E. (1969). Purification of potato aucuba mosaic. *Phytopathology* **59,** 1816–1819.

Kassanis, B. (1956). Serological relationship between potato paracrinkle virus, potato virus S and carnation latent virus. *J. Gen. Microbiol.* **15,** 620–628.

Kassanis, B. (1960). Potato virus M and paracrinkle. *Nature (London)* **188,** 688.

Kassanis, B. (1961). Potato paracrinkle virus. *Eur. Potato J.* **4,** 13–24.

Keeler, R. F., Douglas, D. R., and Stallknecht, G. F. (1975). The testing of blight, aged, and control Russet Burbank potato tuber preparation for ability to produce *spina bifida* and anencephaly in rats, rabbits, hamsters and mice. *Am. Potato J.* **52:** 125–132.

Kehr, A. E., Akeley, R. V., and Houghland, G. V. C. (1964). Commercial potato production. *U.S., Dep. Agric., Handb.* **267.**

Kelman, A. (1953). The bacterial wilt caused by *Pseudomonas solanacearum*. *N.C. Agric. Exp. Stn., Tech. Bull.* **99.**

Kimpinski, J. (1979). Root lesion nematodes in potatoes. *Am. Potato J.* **56,** 79–86.

Klein, M., Zimmerman-Gries, S., and Sneh, B. (1976). Association of bacterialike organisms with a new potato disease. *Phytopathology* **66,** 564–569.

Knorr, L. (1945). Reliability of the stem-ooze test for field identification of potato ring rot. *Am. Potato J.* **22,** 57–62.

Koch, K. L. (1933). The nature of potato rugose mosaic. *Phytopathology* **23**, 319–342.

Kohler, E. (1948). Ein Schnellverfahren zum Nachweis des Kartoffel-A Virus (vorlaufige Mitteilung). [A rapid method for the demonstration of potato virus A (preliminary note).] *Kartoffelwirtsch* **1** (3), 56.

Kohler, E. (1960). Inoculation experiments with potato virus A (str. 'Magna 556') on *N. glutinosa* and some other hosts. *Zentralbl. Bakteriol., Parasitenkd., Abt. 2, Infektionskr. Hyg., Naturwiss.: Allg., Landwirtsch. Tech. Mikrobiol.* **113**, 19–21.

Korf, R. P., and Dumont, K. P. (1972). Whetzelinia, a new generic name for Sclerotinia and *S. tuberosa. Mycologia* **64**, 248–251.

Kowalska, A. (1977). Reaction of red kidney bean to potato virus S. *Potato Res.* **20**, 85–88.

Kowalska, A., and Waś, M. (1976). Detection of potato virus M and potato virus S on test plants. *Potato Res.* **19**, 131–139.

Kraus, J. E. (1945). Influence of certain factors on second growth on Russet Burbank potatoes. *Am. Potato J.* **22**, 134–142.

Krause, R. A., Massie, L. B., and Hyre, R. A. (1975). Blitecast: A computerized forecast of potato late blight. *Plant Dis. Rep.* **59**, 95–98.

Ksiażek, D. (1975). Z badán nad reakcja niektórych chwastów na wirusy ziemniaka M, S i Y. [Studies on the reaction of some weeds to potato viruses M, S and Y.] *Postepy Nauk Roln.* **22/27** (6), 29–32; (Pl) *Inst. Ekol, PAN, Dziekanów Leśny,* Poland. *Rev. Plant Pathol.* **55**, 842.

Kunkel, R. (1972). "Cats eye" in Kennebec potatoes. *Am. Potato J.* **42**, 357.

Kunkel, R., and Dow, A. I. (1961). A possible function of potassium in decreasing susceptibility of Russet Burbank potatoes to black spot when bruised. *Am. Potato J.* **38**, 368–369.

Kunkel, R., and Gardner, W. H. (1959). Black spot of Russet Burbank potatoes. *Proc. Am. Soc. Hortic. Sci.* **73**, 436–444.

Kunkel, R., and Gardner, W. H. (1965). Potato tuber hydration and its effect on black spot of Russet Burbank potatoes in the Columbia Basin of Washington. *Am. Potato J.* **42**, 109–124.

Kunkel, R., and Weller, M. (1966). Fumigation of potato soils in Washington. *Am. Potato J.* **42**, 57–69.

Langerfeld, E. (1973). Effect of temperature on infection of potato tubers by fungi of the genus *Fusarium* Lk. *Potato Res.* **16**, 224–233.

Lapwood, D. H., and Dyson, P. W. (1966). An effect of nitrogen on the formation of potato tubers and the incidence of common scab (*Streptomyces scabies*). *Plant Pathol.* **15**, 9–14.

Lapwood, D. H., and Hide, G. A. (1971). Potato diseases. *In* "Diseases of Crop Plants" (J. H. Western, ed.), pp. 89–122. Wiley, New York.

Lapwood, D. H., Hide, G. A., and Lennard, J. H. (1977). Effects of cultural practices, *Verticillium nubilum* and locality on coiled sprout and yield of early potatoes. *Ann. Appl. Biol.* **85**, 203–215.

Large, E. C. (1940). "The Advance of the Fungi." Holt, New York.

Larsen, E. C. (1949). Investigations on cause and prevention of greening of potato tubers. *Idaho, Agric. Exp. Stn., Bull.* **16**.

Larson, R. H., and Albert, A. R. (1945). Physiological internal necrosis of potato tubers in Wisconsin. *J. Agric. Res.* **71**, 487–505.

Larson, R. H., and Albert, A. R. (1949). Relation of potato varieties to incidence of physiological internal tuber necrosis. *Am. Potato J.* **26**, 427–431.

Leach, J. G., and Bishop, C. F. (1946). Purple-top wilt (blue stem) of potatoes. *Bull.—W.Va., Agric. Exp. Stn.* **326**.

Leach, S. S. (1975). Control of Fusarium tuber dry rot of white potatoes. *U.S., Agric. Res. Serv., Northeast. Reg. [Rep.] ARS-NE* **ARS-NE-55.**

Leach, S. S. (1976). Increasing storability of potatoes by chemical treatments at time of storage. *Am. Potato J.* **53,** 364.

Leach, S. S., and Nielsen, L. W. (1975). Elimination of fusarial contamination on seed potatoes. *Am. Potato J.* **52,** 211–218.

Leach, S. S., and Rich, A. E. (1969). The possible role of parasexuality and cytoplasmic variation in race differentiation in *Phytophthora infestans*. *Phytopathology* **59,** 1360–1365.

Leach, S. S., and Webb, R. E. (1975). Screening for resistance to Fusarium tuber rot. *Am. Potato J.* **52,** 246.

LeClerg, E. L., Lombard, P. M., Eddins, A. H., Cook, H. T., and Campbell, J. C. (1944). Effect of different amounts of spindle tuber and leaf roll on yields of Irish potatoes. *Am. Potato J.* **21,** 60–71.

Lee, S. T. (1972). Occurrence of potato virus M and its strains in Taiwan. *Proc. Natl. Sci. Counc. Repub. China* **5,** 91–108.

Lee, S. T. (1976). Relationships of strains of potato virus A in Taiwan. *J. Agric. Assoc. China* **93,** 60–71.

Levitt, J. (1942). A histological study of hollow heart of potatoes. *Am. Potato J.* **19,** 134–143.

Lihnell, D. (1959). Studies on the etiology of spraing. *Am. Potato J.* **36,** 299–300.

Liljemark, A., and Widoff, E. (1960). Greening and solanine development of white potato in fluorescent light. *Am. Potato J.* **37,** 377–389.

Locke, S. B. (1947). Field resistance to leafroll infection in potato varieties. *Phytopathology* **37,** 14.

Logan, C., Copeland, R. B., and Little G. (1975). Potato gangrene control by ultra low volume sprays of thiabendazole. *Ann. Appl. Biol.* **80,** 199–204.

Lutman, B. F. (1919). Tipburn of the potato and other plants. *Bull.—Vt., Agric. Exp. Stn.* **214.**

Lutman, B. F. (1922). The relation of the water pores and stomata of the potato leaf to the early stages and advance of tipburn. *Phytopathology* **12,** 305–333.

Lutman, B. F., and Wheeler, H. E. (1948). *Bacillus Megaterium* deBary from the interior of healthy potato tubers. *J. Wash. Acad. Sci.* **38,** 336–340.

Lutz, J. M., Findlen, H., and Ramsey, G. B. (1951). Quality of Red River Valley potatoes in various types of consumer packages. *Am. Potato J.* **28,** 589–602.

McClements, W. L., and Kaesberg, P. (1977). Size and secondary structure of potato spindle tuber viroid. *Virology* **76,** 477–484.

McCrum, R. C., and Manzer, F. E. (1967). Serological comparison of *Streptomyces* spp. isolated from potato. *Am. Potato J.* **44,** 338.

MacDonald, D. M. (1973). Heat treatment and meristem culture as a means of freeing potato from viruses X and S. *Potato Res.* **16,** 263–269.

MacGarvie, Q. D., and Hide, G. A. (1966). *Verticillium* species from potato seed stocks in Britain in 1965. *Plant Pathol.* **15,** 72–75.

McKay, M. B. (1926). Potato wilt and its control. *Oreg., Agr. Exp. Stn., Bull.* **221.**

McKee, R. K. (1968). Effect of soil moisture on incidence of potato scab. *Eur. Potato J.* **11,** 111–116.

MacKinnon, J. P. (1970). Comparative levels of leafroll virus resistance in potato varieties and seedlings. *Am. Potato J.* **47,** 444–446.

MacLachlan, D. S. (1960). Potato spindle tuber in Eastern Canada. *Am. Potato J.* **37,** 13–17.

MacLachlan, D. S., Larson, R. H., and Walker, J. C. (1953). Strain interelationships in potato virus A. *Res. Bull.—Wis., Agric. Exp. Stn.* **180.**

McLean, J. G. (1952). Results of testing lines and varieties of potatoes for field resistance to Verticillium wilt. *Phytopathology* **42**, 26 (Abstr.).

McLean, J. G., and Akeley, R. V. (1957). "Control of Verticillium Wilt," Potato Handb. No. 2. Potato Assoc. Am., New Brunswick, New Jersey.

Maat, D. Z., and Bokx, J. A. de. (1978). Potato leafroll virus: antiserum preparation and detection in potato leaves and sprouts with the enzyme-linked immunosorbent assay (ELISA). *Netherlands J. of Pl. Pathol.* **84**, 149–156.

Mai, W. F. (1947). Virus X in the new potato varieties and the transmission of the virus by the cutting knife. *Am. Potato J.* **24**, 341–351.

Mai, W. F., and Lear, B. (1953). The golden nematode. *Cornell Exten. Bull.* **870**.

Mai, W. F., and Spears, J. F. (1954). The golden nematode in the United States. *Am. Potato J.* **31**, 387–396.

Mai, W. F., Crittenden, H. W., and Jenkins, W. R. (1960). Distribution of stylet-bearing nematodes in the northeastern United States. *Bull.—N.J., Agric. Exp. Stn.* **667**.

Mai, W. F., Bloom, J. R., and Chen, T. A. (1977). Biology and ecology of the plant-parasitic nematode *Pratylenchus penetrans. Bull—Pa., Agric. Exp. Stn.* **815**.

Malcolmson, J. F. (1959). A study of Erwinia isolates obtained from soft rots and blackleg of potatoes. *Trans. Br. Mycol. Soc.* **42**, 261–269.

Malcolmson, J. F. (1960). A disease of potatoes and tomatoes caused by *Bacillus subtilis. Rep. Scot. Soc. Res. Plant Breed.* pp. 24–28.

Malcolmson, J. F., and Gray, E. G. (1968a). Factors affecting the occurrence of gangrene *(Phoma exigua)* in potatoes. *Ann. Appl biol.* **62**, 77–87.

Malcolmson, J. F., and Gray, E. G. (1968b). The incidence of gangrene of potatoes caused by *Phoma exigua* in relation to handling and storages. *Ann. Appl. Biol.* **62**, 89–101.

Malcolmson, J. F., and Gray, E. G. (1968c). A note on fungi assigned to *Phoma exigua* with special reference to those causing gangrene of potatoes. *Trans. Br. Mycol. Soc.* **51**, 618–620.

Manzer, F. E., and Merriam, D. (1961). Field transmission of the potato spindle tuber virus and virus X by cultivating and hilling equipment. *Am. Potato J.* **38**, 346–352.

Manzer, F. E., and Merriam, D. C. (1974). Importance of overwintered early blight-infected potato vines in Maine. *Am. Potato J.* **51**, 419–420.

Manzer, F. E., Akeley, R. V., and Merriam, D. (1964a). Resistance to powdery scab in *Solanum tuberosum* L. *Am. Potato J.* **41**, 374–376.

Manzer, F. E., Akeley, R. V., and Merriam, D. (1964b). Resistance in *Solanum tuberosum* to mechanical inoculation with potato spindle tuber virus. *Am. Potato J.* **41**, 411–416.

Manzer, F. E., McCrum, R. C., and Merriam D. C. (1975). Testing Maine potato seed stocks in Florida for presence of virus X. *Am. Potato J.* **52**, 137–141.

Manzer, F. E., McIntyre, G. A., and Merriam, D. C. (1977). A new potato scab problem in Maine. *Maine, Agric. Exp. Stn., Tech. Bull.* **85**.

Manzer, F. E., Merriam, D. C., Storch, R. H., and Simpson, G. W. (1982). Effect of time of inoculaton with potato leafroll virus on development of net necrosis and stem-end browning in potato tubers. *Am. Potato J.* **59**, 337–349.

Maramorosch, K., Granados, R. R., and Hirumi, H. (1970). Mycoplasma diseases of plants and insects. *Adv. Virus Res.* **16**, 135–193.

Martyn, E. B. (1968). Plant virus names. An annotated list of names of plant viruses and diseases. *Commonw. Mycol. Inst. Phytopathol. Paper* **9**, 1–204.

Matthews, R. E. F. (1970). "Principles of Plant Virology." Academic Press, New York.

Maurer, A. R., Van Adrichem, M., Young, D. A., and Davies, H. T. (1968). Cariboo, a new late potato variety of distinctive appearance. *Am. Potato J.* **45**, 247–249.

Mendoza, H. A., and Haynes, F. L., Jr. (1974). Early detection of virus X (PVX) in potato tubers. *Am. Potato J.* **51**, 306.

Menzies, J. D. (1950a). *Erysiphe cichoracearum* DC as a parasite of potatoes. *Plant Dis. Rep.* **34**, 140–141.

Menzies, J. D. (1950b). Purple-top-type viruses of potatoes in Washington. *Phytopathology* **40**, 968 (Abstr.).

Menzies, J. D. (1957). Control of potato scab by a scab-suppressing factor in certain soils. *Phytopathology* **47**, 528 (Abstr.).

Menzies, J. D. (1959). Occurrence and transfer of a biological factor in the soil that suppresses potato scab. *Phytopathology* **49**, 648–652.

Menzies, J. D., and Giddings, N. J. (1953). Identity of potato curly top and green dwarf. *Phytopathology* **43**, 684–686.

Merriam, D. C., and Akeley, R. V. (1974). A new local lesion host for virus. X. *Am. Potato J.* **51**, 305 (Abstr.).

Milbrath, J. A. (1946). Green dwarf: A virus disease of potato. *Phytopathology* **36**, 671–674.

Millard, W. A., and Burr, S. (1926). A study of twenty-four strains of *Actinomyces* and their relation to types of common scab of potato. *Ann. Appl. Biol.* **13**, 580–644.

Miller, P. M., and Edgington, L. V. (1962). Controlling parasitic nematodes and soil-borne diseases with soil fumigation. *Am. Potato J.* **39**, 235–240.

Miller, P. M., and Hawkins, A. (1969). Long term effects of preplant fumigation of potato fields. *Am. Potato J.* **46**, 387–397.

Miller, P. M., Edgington, L. V., and Hawkins, A. (1967). Effects of soil fumigation on Verticillium wilt, nematodes and other diseases of potato roots and tubers. *Am. Potato J.* **44**, 316–323.

Miller, P. R., and Pollard, H. L. (1976). "Multilingual Compendium of Plant Diseases." Am. Phytopathol. Soc., St. Paul, Minnesota.

Miller, P. R., and Pollard, H. L. (1977). "Multilingual Compendium of Plant Diseases," Vol. II. Phytopathol. Soc., St. Paul, Minnesota.

Mills, W. R. (1964). Pennchip, a new potato variety resistant to late blight and scab with superior chipping quality. *Am. Potato J.* **41**, 54–58.

Miska, J. P., and Nelson, G. A. (1975). Potato seed-piece decay: A bibliography, 1930–1975. *Can. Plant Dis. Surv.* **55**, 126–146.

Molina, J. J., Harrison, M. D., and Brewer, J. W. (1974). Transmission of *Erwinia carotovora* var. *atroseptica* by *Drosophila melanogaster* Meig. I. Acquisition and transmission of the bacterium. *Am. Potato J.* **51**, 251–253.

Morel, G., Martin, C., and Muller, J. F. (1968). La guérison des Pommes de Terre atteintes de maladies a virus. [Production of virus-free potatoes.] *Ann. Physiol. Veg.* **10**, 113–139.

Morsink, F. (1966). Interaction of *Pratylenchus penetrans* and *Verticillium albo-atrum* in the Verticillium wilt of potatoes, and attraction of *Pratylenchus penetrans* by various chemicals. Ph.D. Thesis, University of New Hampshire, Durham.

Morsink, F., and Rich, A. E. (1968). Interactions between *Verticillium albo-atrum* and *Pratylenchus penetrans* in the Verticillium wilt of potatoes. *Phytopathology* **58**, 401 (Abstr.).

Mortvedt, J. J., Fleischfresser, M. H., Berger, K. C., and Darling, H. M. (1961). The relation of soluble manganese to the incidence of common scab in potatoes. *Am. Potato J.* **38**, 95–100.

Mulvey, R. H., and Stone, A. R. (1976). Description of *Punctodera matadorensis* n. gen., n. sp. (Nematoda: Heteroderidae) from Saskatchewan with lists of species and generic diagnosis of Globodera (n. rank), Heterodera, and Sarisodera. *Can J. Zool.* **54**, 772–775.

Muncie, J. H. (1949). Reaction of hybrid potato varieties to infection by *Fusarium eumartii*. *Am. Potato J.* **26**, 100 (Abstr.).

Munro, J. (1961). The importance of potato virus X. *Am. Potato J.* **38**, 440–447.

Munro, J. (1975). "Seed Potatoes from Canada." Dept. of Industry, Trade and Commerce, Ottawa.

Munro, J. (1978). Seed potato improvement in Canada. *Can. Plant Dis. Surv.* **58**, 26–28.

Murphy, H. J., Goven, M. J., and Merriam, D. C. (1966). Effect of three viruses on yield, specific gravity and chip color of potatoes in Maine. *Am. Potato J.* **43**, 393–396.

Murphy, H. J., Morrow, L. S., Young, D. A., Ashley, R. A., Orzolek, M. D., Precheur, R. J., Wells, O. S., Jensen, R., Henninger, M. R., Sieczka, J. B., Pisarczyk, J. S., Cole, R. E., Wakefield, R. E., and Young, R. J. (1982). Performance evaluations of potato clones and varieties in the northeastern states 1981. *Maine Life Sci. and Agr. Exp. Sta. Bull.* **782.**

Nagaich, B. B., and Giri, B. K. (1973). Purple top roll disease of potato. *Am. Potato J.* **50**, 79–85.

Natti, J. J., Kirkpatrick, H. C., and Ross, A. F. (1953). Host range of potato leafroll virus. *Am. Potato J.* **30**, 55–64.

Nelson, G. A., and Torfason, W. E. (1974). Association effects of leaf roll and ring rot on disease expression and yield of potatoes. *Am. Potato J.* **51**, 12–15.

Newton, W., and Lines, C. (1947). The dusting of cut potato tubers as a preventive against Pythium rot. *Sci. Agric.* **27**, 72–73.

Niederhauser, J. S., and Cervantes, J. (1959). Anita, Bertita, and Conchita, three new blight-resistant potato varieties developed in central Mexico. *Am. Potato J.* **36**, 300–301.

Niederhauser, J. S., Buck, R. W., and Akeley, R. V. (1959). Erendira, a new blight-resistant potato variety for the highlands of central Mexico. *Am. Potato J.* **36**, 300 (Abstr.).

Nielsen, L. W. (1954). The susceptibility of seven potato varieties to bruising and bacterial soft rot. *Phytopathology* **44**, 30–35.

Nielsen, L. W. (1968). Accumulation of respiratory CO_2 around potato tubers in relation to bacterial soft rot. *Am. Potato J.* **45**, 174–181.

Nielsen, L. W., and Haynes, F. L. (1960). Resistance in *Solanum tuberosum* to *Pseudomonas solanacearum*. *Am. Potato J.* **37**, 260–267.

Nienhaus, F. (1960). Test der Mosaikviren Y, X, und A unmittelbar von der Kartoffelknolle. [Test of the mosaic viruses Y, X, and A directly from the potato tuber.] *Naturwissenschaften* **47**, 164–165.

Nonnecke, I. L., Torfason, W. E., and Young, L. C. (1966). Chinook: A new high quality potato resistant to common scab. *Am. Potato J.* **43**, 1–5.

O'Brien, M. J. (1972). Hosts of potato spindle tuber virus in suborder Solanineae. *Am. Potato J.* **49**, 70–72.

O'Brien, M. J., and Raymer, W. B. (1964). Symptomless hosts of the potato spindle tuber virus. *Phytopathology* **54**, 1045–1047.

O'Brien, M. J., and Rich, A. E. (1976). "Potato Diseases." *U.S., Dep. Agric., Handb.* **474.**

O'Brien, M. J., and Thirumalachar, M. J. (1972). The identity of the potato smut. *Sydowia* **26**, 199–203.

O'Brien, M. J., and Thirumalachar, M. J. (1977). Identity of the fungi inciting charcoal rot disease. *Sydowia* **30**, 141–144.

Ohms, R. E., and Fenwick, H. S. (1961). Potato early blight—symptoms, cause and control. *Idaho, Agric. Exten. Serv., Bull.* **346.**

O'Keefe, R. B. (1970a). Shurchip: A new scab resistant processing potato variety. *Am. Potato J.* **47**, 124–129.

O'Keefe, R. B. (1970b). Sioux: A scab resistant red potato variety. *Am. Potato J.* **47**, 163–168.

O'Keefe, R. B., and Werner, H. O. (1965). Platte and Hi-Plains, two new scab resistant processing varieties. *Am. Potato J.* **42**, 361–368.

Orton, W. A. (1914). Potato wilt, leaf roll, and related diseases. *U.S., Dep. Agric., Bur. Plant Ind. Bull.* **64.**

Oswald, J. W. (1950). A strain of the alfalfa mosaic virus causing vine and tuber necrosis in potato. *Phytopathology* **40**, 973–991.

Oswald, J. W., and Lorenz, O. A. (1956). Soybeans as a green manure crop for the prevention of potato scab. *Phytopathology* **46**, 22 (Abstr.).

Oswald, J. W., and Lorenz, O. A. (1957). Potassium and internal black spot of potato in California. *Phytopathology* **47**, 530 (Abstr.).

Parker, M. M., Akeley, R. V., and Stevenson, F. J. (1954). Pungo: A new variety of potato resistant to late blight and adapted to eastern Virginia. *Am. Potato J.* **31**, 322–326.

Pavek, J. J., Douglas, D. R., McKay, H. C., and Ohms, R. E. (1973a). Targhee: An oblong russet potato variety with attractive tubers and high resistance to common scab. *Am. Potato J.* **50**, 293–295.

Pavek, J. J., Douglas, D. R., McKay, H. C., and Ohms, R. E. (1973b). Nampa: A long russet potato variety with tolerance to high daytime temperatures. *Am. Potato J.* **50**, 296–299.

Peterson, C. E., and Hooker, W. J. (1958). Tawa: A new early potato variety resistant to late blight, common scab and immune to latent mosaic. *Am. Potato J.* **36**, 267–274.

Peterson, C. E., Ellis, N. K., Akeley, R. V., and Stevenson, F. J. (1954). Cherokee: A new medium maturing potato variety resistant to common scab, late blight, mild mosaic and net necrosis. *Am. Potato J.* **31**, 53–57.

Peterson, L. C. (1957). "Control of Late Blight," Potato Handb. No. 2. Potato Assoc. Am., New Brunswick, New Jersey.

Peterson, L. C., and Plaisted, R. L. (1966). Peconic, a new variety resistant to the golden nematode. *Am. Potato J.* **43**, 450–452.

Plaisted, R. L., and Peterson, L. C. (1972). Inheritance of the Katahdin "knobby tuber." *Am. Potato J.* **49**, 285.

Plaisted, R. L., Thurston, H. D., Peterson, L. C., Fricke, D. H., Cetas, R. C., Harrison, M. B., Sieczka, J. B., and Jones, E. D. (1973). Hudson: a high yielding variety resistant to golden nematode. *Am. Potato J.* **50**, 212–215.

Polli, J., and Moeller, S. (1944). Trials for the control of powdery mildew on potatoes. *Palest. J Bot., Rehovot Ser.* **4**, 148–156.

Pond, D. D. (1964). Field control of potato leafroll virus with systemic insecticides. *Am. Potato J.* **41**, 14–17.

Potter, H. S. (1981). Fungigation for the control of potato diseases. *Am. Potato J.* **58**, 514–515.

Potter, H. S., and Hooker, W. J. (1969). Low gallonage fungicide spraying for potatoes. *Am. Potato J.* **46**, 434 (Abstr.).

Powelson, R. L., and Carter, G. E. (1973). Efficacy of soil fumigants for the control of Verticillium wilt of potatoes. *Am. Potato J.* **50**, 162–167.

Pratt, A. J. (1969). Pride: A new early potato variety with shallow eyes and some scab resistance. *Am. Potato J.* **46**, 88–90.

Racicot, H. N., Savile, D. B. O., and Conners, I. L. (1938). Bacterial wilt and rot of potatoes—Some suggestions for its detection, verification, and control. *Am. Potato J.* **15**, 312–318.

Radewald, J. D., Harvey, O. A., Shibuya, F., and Nelson, J. (1975). A progress report on the control of the root-knot nematode on White Rose potatoes with granular nematicides. *Calif. Agric.* **29**, 8–9.

Ramsey, G. B. (1941). Botrytis and Sclerotinia as potato tuber pathogens. *Phytopathology* **31**, 439–448.

Raymer, W. B. (1975). Potato spindle tuber viroid (PSTV)–tomato assay. *Am. Potato J.* **52**, 242–243.

Raymer, W. B., and Milbrath, J. A. (1957). A local-lesion test for potato virus A in the presence of potato virus X. *Phytopathology* **47**, 532 (Abstr.).

Raymer, W. B., and O'Brien, M. J. (1962). Transmission of potato spindle tuber virus to tomato. *Phytopathology* **52**, 749 (Abstr.).

Rebois, R. V., and Webb, R. E. (1979). Reniform nematode resistance in potato clones. *Am. Potato J.* **56**, 313–319.

Rebois, R. V., Eldridge, B. J., and Webb, R. E. (1978). *Rotylenchus reniformis* parasitism of potatoes and its effect on yields. *Plant Dis. Rep.* **62**, 520–523.

Rhodes, J. A., and Van Rooyen, C. E. (1968). "Textbook of virology; for students and practitioners of medicine and the other health sciences." Williams & Wilkins, Baltimore, Maryland.

Rich, A. E. (1949). Varietal resistance to potato late blight in western Washington in 1948. *Plant Dis. Rep.* **33**, 11.

Rich, A. E. (1950). The effect of various defoliants on potato vines and tubers in Washington, 1948. *Am. Potato J.* **27**, 87–92.

Rich, A. E. (1951). Phloem necrosis of Irish potatoes in Washington. *Bull.—Wash., Agric. Exp. Stn.* **528**.

Rich, A. E. (1968). Potato diseases. *In* "Potatoes: Production, Storing, Processing" (O. Smith), pp. 397–437. Avi Publ., Westport, Connecticut.

Rich, A. E. (1969). Inactivation of potato virus X in Green Mountain potatoes. *Phytopathology* **59**, 710–711.

Rich, A. E. (1977). Potato diseases. *In* "Potatoes: Production, Storing, Processing" (O. Smith), 2nd ed., pp. 506–549. Avi Publ., Westport, Connecticut.

Rich, A. E., Campbell, C. S., Mullany, R. E., Manzer, F. E., and Blood, P. T. (1960). The influence of various fungicides and antibiotics on cut seed potatoes. *Am. Potato J.* **37**, 351.

Rich, S., and Hawkins, A. (1970). The susceptibility of potato varieties to air pollutants in the field. *Phytopathology* **60**, 1309.

Richardson, L. T., and Phillips, W. R. (1949). Low temperature breakdown of potatoes in storage. *Sci. Agric.* **29**, 149–166.

Riedl, W. A. (1968). Wyred: A new high yielding red potato variety. *Am. Potato J.* **45**, 33–35.

Riedl, W. A., Stevenson, F. J., and Bonde, R. (1946). The Teton potato: A new variety resistant to ring rot. *Am. Potato J.* **23**, 379–389.

Rieman, G. H. (1962). Superior: A new white, medium-maturing scab-resistant potato variety with high chipping quality. *Am. Potato J.* **39**, 19–28.

Rieman, G. H., and McFarlane, J. S. (1943). The resistance of the Sebago variety to yellow dwarf. *Am. Potato J.* **20**, 277–283.

Rieman, G. H., and Schultz, J. H. (1955). Red Beauty: A new bright-red medium-maturing variety of potato, resistant to Verticillium wilt. *Am. Potato J.* **32**, 346–349.

Rieman, G. H., and Young, D. A. (1955). Antigo: A new white medium-maturing potato variety resistant to common scab. *Am. Potato J.* **32**, 407–410.

Roberts, D. A., and Boothroyd, C. W. (1972). "Fundamentals of Plant Pathology." Freeman, San Francisco, California.

Roberts, F. M. (1946). Underground spread of potato virus X. *Nature (London)* **48**, 663.

Robinson, D. B., and Ayers, G. W. (1953). The control of Verticillium wilt of potatoes by seed treatment. *Can. J. Agric. Sci.* **33**, 147–152.

Robinson, D. B., and Ayers, G. W. (1961). Verticillium wilt of potato in relation to vascular infection of the tuber. *Canad. J. Pl. Sci.* **41**, 703–708.

Robinson, D. B., and Callbeck, L. C. (1955). Stem streak necrosis of potato in Prince Edward Island. *Am. Potato J.* **32**, 418–423.

Robinson, D. B., Larson, R. H., and Walker, J. C. (1957). Verticillium wilt of potato in relation to symptoms, epidemiology, and variability of the pathogen. *Res. Bull.—Wis., Agric. Exp. Stn.* **202**.

Robinson, D. B., Easton, G. D., and Larson, R. H. (1960). Some common stem streaks of potato. *Am. Potato J.* **37**, 67–72.

Rose, D. H., and Fisher, D. F. (1940). Desiccation effects on skinned potatoes. *Am. Potato J.* **17**, 287–289.

Ross, A. F. (1948). Local lesions with virus Y. *Phytopathology* **38**, 930–932.

Ross, H. (1968). *Lycopersicum chilense* Dun., eine Testpflanze für die beiden Kartoffelviren M and S. [*Lycopersicon chilense* a test plant for the two potato viruses M and S.] *Eur. Potato J.* **11**, 281–286.

Rotem, J. (1959). The influence of sandstorms in the Negev on the sensitivity of potatoes and tomatoes to the early blight disease. *Bull. Res. Counc. Isr. Sect. D,* 100–102.

Rowe, R. C. (1975). Powdery mildew of potatoes in Ohio. *Plant Dis. Rep.* **59**, 330–331.

Rowe, R. C., and Schmitthenner, A. F. (1977). Potato pink rot in Ohio caused by *Phytophthora erythroseptica* and *P. cryptogea. Plant Dis. Rep.* **61**, 807–810.

Ruehle, G. D. (1940). Bacterial soft rot of potatoes in southern Florida. *Bull.—Fla., Agric. Exp. Stn. Tech. Bull.* **348.**

Salt, G. A. (1964). The incidence of *Oospora pustulans* on potato plants in different soils. *Plant Pathol.* **13**, 155–158.

Saltykova, L. P. (1973). O sortovykh razlichiyakh kartofelya po ustoĭchivosti k obychnoĭ i agressivnym rasam vozbuditelya raka. [Varietal differences in resistance of potato to ordinary and aggressive races of the causal agent of wart.] *Tr. Vses. Nauchno-Issled. Inst. Zashch. Rast.* **36**, 43–62, *Rev. Plant Pathol.* **56**, 83.

Sand, P. F. (1979). The importance of potato varieties resistant to golden nematode in the cooperative federal–state golden nematode program. *Am. Potato J.* **56**, 478 (Abstr.).

Sanford, L. L., Iritani, W. M., McLean, J. G., Akeley, R. V., and Sparks, W. C. (1964). Shoshoni: A new russet-skinned potato with resistance to common scab and Verticillium wilt. *Am. Potato J.* **41**, 95–99.

Sasser, J. N. (1954). Indentification and host-parasite relationships of certain root-knot nematodes *(Meloidogyne* spp.). *Md., Agric. Exp. Stn., Tech. Bull.* **A-77.**

Sawyer, R. L. (1958). Black spot symposium summary. *Am. Potato J.* **36**, 76–78.

Sawyer, R. L., and Collin, G. H. (1960). Black spot of potatoes. *Am. Potato J.* **37**, 115–126.

Schaal, L. A., and Livingston, C. H. (1958). Experiments on the use of pentachloronitrobenzene (PCNB) as a control for potato scab and Rhizoctonia. *Am. Potato J.* **35**, 445–446.

Schaal, L. A., Edmundson, W. C., and Kunkel, R. (1949). Yampa: A new scab-resistant potato. *Am. Potato J.* **26**, 335–342.

Scholey, J., Marshall, C., and Whitbread, R. (1968). A pathological problem associated with prepackaging of potato tubers. *Plant Pathol.* **17**, 135–139.

Schultz, E. S. (1919). Investigations of the mosaic disease of the Irish potato. *J. Agric. Res.* **17**, 247–273.

Schultz, E. S., and Folsom, D. (1920). Transmission of the mosaic disease of Irish potatoes. *J. Agric. Res.* **19**, 315–338.

Schultz, E. S., and Folsom, D. (1923). A spindling tuber disease of Irish potatoes. *Science* **57**, 149.

Schultz, E. S., and Folsom, D. (1925). Transmission, variation, and control of certain degeneration diseases of Irish potatoes. *J. Agric. Res.* **25**, 43–117.

Schultz, E. S., Stevenson, F. J., and Akeley, R. V. (1947). Resistance of potato to virus Y, the cause of veinbanding mosaic. *Am. Potato J.* **24**, 413–419.

Schultz, T. H., Berger, K. C., Darling, H. M., and Fleischfresser, M. H. (1961). Urea formaldehyde concentrate-85 for scab control in potatoes. *Am. Potato J.* **38**, 85–88.

Schulz, J. T. (1963). *Tetranychus telarius* (L.) new vector of virus Y. *Plant Dis. Rep.* **47**, 594–596.

Schulz, J.T. (1976). Insects infesting potatoes in the Red River Valley. *In* "Potato Production in North Dakota." *North Dakota State Univ. Agric. and Appl. Sci. Ext. Bull.* **26**, 50–52.

Schumann, G. L., Thurston, H. D., and Plaisted, R. L. (1977). Potato spindle tuber viroid—Reducing the threat to new varieties. *N.Y. Food Life Sci. Q.* **11**(3), 20–23.

Schumann, G. L., Tingey, W. M., and Thurston, H. D. (1980). Evaluation of six insect pests for transmission of potato spindle tuber viroid. *Am. Potato J.* **57**, 205–211.

Scott, J. D. (1976). Praise the potato! *Reader's Digest,* Dec., pp. 205–212.

Shands, W. A., and Simpson, G. W. (1969). Bioenvironmental control of the green peach aphid, *Myzus persicae. Am. Potato J.* **46**, 56–58.

Shands, W. A., and Simpson, G. W. (1971). Seasonal history of the buckthorn aphid and suitability of alder-leaved buckthorn as a primary host in northeastern Maine. *Maine, Agric. Exp. Stn., Tech. Bull.* **51.**

Shands, W. A., Simpson, G. W., Seaman, B. A., Roberts, F. S., and Flynn, C. M. (1972). Effects of differing abundance levels of aphids and of certain virus diseases upon yield and virus disease spread of potatoes. *Maine, Agric. Exp. Stn., Tech. Bull.* **56.**

Shenider, Yu. I., and Murzakova, K. F. (1964). Bakteriozy Kartofelya. [Bacteriosis of potato.] *Zashch. Rast. Vred. Bolean. (Minist. Sel'sk. Khoz SSSR)* **9**, 32–33, *Rev. Appl. Mycol.* **44**, 103.

Shuja, M. A. (1969). Effect of dates of planting, varieties and spacings on occurrence of internal brown spot in potatoes. *West Pak. J. Agric. Res.* **7**, 32–37.

Simpson, G. W. (1977). Potato insects and their control. *In* "Potatoes: Production, Storing, Processing" (O. Smith, ed.), 2nd ed., pp. 550–605. Avi Publ., Westport, Connecticut.

Simpson, G. W., and Akeley, R. V. (1964). Penobscot: A new variety of potato with leafroll resistance and high solids. *Am. Potato J.* **41**, 140–144.

Simpson, G. W., and Shands, W. A. (1949). Progress on some important insect and disease problems of Irish potato production in Maine. *Maine, Agric. Exp. Stn., Bull.* **470.**

Singh, R. P. (1970). Occurrence, symptoms, and diagnostic hosts of strains of potato spindle tuber virus. *Phytopathology* **60**, 1314 (Abstr.).

Singh, R. P. (1973). Experimental host range of the potato spindle tuber virus. *Am. Potato J.* **50**, 111–123.

Singh, R. P., and Bagnall, R. H. (1968). *Solanum rostratum* Duval., a new test plant for the potato spindle tuber virus. *Am. Potato J.* **45**, 335–336.

Singh, R. P., and McDonald, J. G. (1981). Purification of potato virus A and its detection in potato by enzyme-linked immunosorbent assay. *Am. Potato J.* **58**, 181–189.

Singh, R. P., Finnie, R. E., and Bagnall, R. H. (1971). Losses due to the potato spindle tuber virus. *Am. Potato J.* **48**, 262–267.

Singh, R. P., Drew, M. E., and Smith, E. M. (1977). Detection of potato virus A by *Physalis floridana* leaf test. *Am. Potato J.* **54**, 479.

Singh, S. (1972). Spinach *(Spinacia oleracea)* a new test plant for potato virus X. *Indian Phytopathol.* **25**, 147–148.

Smith, K. M. (1957). "A Textbook of Plant Virus Diseases," 2nd ed. Little, Brown, Boston, Massachusetts.

Smith, K. M. (1972). "A Textbook of Plant Virus Diseases," 3rd ed. Academic Press, New York.

Smith, M. A., and Ramsey, G. B. (1947). Bacterial lenticel infection of early potatoes. *Phytopathology* **37**, 225–242.

Smith, O. (1968). "Potatoes: Production, Storing, Procesing." Avi Publ., Westport, Connecticut.

Smith, W. L., Jr. (1949). Some specific characters of *Erwinia atroseptica* and *Erwinia carotovora. Am. Potato J.* **39**, 22–23.

Smith, W. L., Jr., and Wilson, J. B. (1978). Market diseases of potatoes. *U.S., Dep. Agric., Handb.* **479.**

Smott, J. J., Gough, F. J., and Gallegly, M. E. (1957). Oospore formation in *Phytophthora infestans. Phytopathology* **47,** 33 (Abstr.).

Snyder, W. C., and Hansen, H. N. (1940). The species concept in *Fusarium. Am. J. Bot.* **27,** 64-67.

Synder, W. C., and Hansen, H. N. (1941). The species concept in *Fusarium* with reference to section *Martiella, Am. J. Bot.* **28,** 738-742.

Snyder, W. C., and Toussoun, T. A. (1965). Current status of taxonomy in *Fusarium* species and their perfect stages. *Phytopathology* **55,** 833-837.

Sogo, J. M., Koller, T., and Diener, T. O. (1973). Potato spindle tuber viroid X. Visualization and size determination by electron microscopy. *Virology* **55,** 70-80.

Soltanpour, P. N., and Harrison, M. D. (1974). Interrelationships between nitrogen and phosphorus fertilization and early blight control of potatoes. *Am. Potato J.* **51,** 1-7.

Stace-Smith, R., and Mellor, F. C. (1968). Eradication of potato viruses X and S by thermotherapy and axillary bud culture. *Phytopathology.* **58,** 290-293.

Stanghellini, M. E., and Meneley, J. C. (1975). Identification of soft-rot Erwinia associated with blackleg of potato in Arizona. *Phytopathology* **65,** 86-87.

Stapp, C. (1947). Neuere Untersuchungen über die Resistenzverschiedenheiten deutscher Kartoffelsorten gegen *Bacterium phytophthorum* Appel. [Recent investigations on the differences in resistance of German potato varieties to *Bacterium phytophthorum* Appel.] *'Festschrift Appel', Biologische Zentralanstalt,* Berlin-Dahlem. *Rev. Appl. Mycol.* **28,** 306-307.

Stevenson, F. J. (1949). Old and new potato varieties. *Am. Potato J.* **26,** 395-404.

Stevenson, F. J., and Clark, C. F. (1938). The Sebago potato: A new variety resistant to late blight. *U.S., Dep. Agric., Circ.* **503.**

Stevenson, F. J., and Livermore, J. R. (1949). The Saranac potato: A new variety promising in Australia. *Am. Potato J.* **26,** 45-46.

Stevenson, F. J., Akeley, R. V., and Brasher, E. P. (1954). Delus: A new variety of potato, immune from the common race of the late blight fungus, high in percentage total solids and adapted to growing conditions in Delaware. *Am. Potato J.* **31,** 410-413.

Stevenson, F. J., McLean, J. G., Hoyman, W. G., and Akeley, R. V. (1955). Early Gem: A new early russet-skin scab-resistant variety of potato adapted to the early potato producing section of Idaho and to certain sections of North Dakota. *Am. Potato J.* **32,** 79-85.

Stevenson, F. J., Akeley, R. V., and Haynes, F. L. (1956). Plymouth: A new variety of potato, immune from the common race of the late blight fungus, moderately resistant to common scab and adapted to growing conditions in the tidelands of North Carolina. *Am. Potato J.* **33,** 296-299.

Stevenson, F. J., Akeley, R. V., Hunter, D., and Beale, W. L. (1970). Seminole: A new variety of potato combining good chipping characteristics with good field performance in Alabama, Florida, and North Carolina. *Am. Potato J.* **47,** 35-38.

Stiles, D. G. (1979). Maine's anti-bruise campaign. *Am. Potato J.* **56,** 481 (Abstr.).

Stobel, G. A., and Rai, P. V. (1968). A rapid serodiagnostic test for potato ring rot. *Plant Dis. Rep.* **52,** 502-504.

Stone, A. R. (1972). *Heterodera pallida* n. sp. (Nematoda: Heteroderidae), a second species of potato cyst nematode. *Nematologica* **18,** 591-596.

Struckmeyer, B. E., and Berger, K. C. (1950). Histological structure of potato stems and leaves as influenced by manganese toxicity. *Plant Physiol.* **25,** 114-119.

Susnoschi, M., Krikun, J., and Zula, Z. (1975). Variety trial of potatoes resistant to Verticillium wilt. *Am. Potato J.* **52,** 227-231.

Sykes, G. B. (1975). The effect of phorate, a wide-spectrum pesticide, on the occurrence of spraing in 12 cultivars of potato. *Plant Pathol.* **24**, 71–73.

Talburt, W. F., and Smith, O. (1975). "Potato Processing," 3rd ed. Avi Publ., Westport, Connecticut.

Tanii, A., and Akai, J. (1975). Blackleg of potato plant caused by a serologically specific strain of *Erwinia carotovora* var. *carotovora* (Jones) Dye. *Ann. Phytopathol. Soc. Jpn.* **41**, 513–517, Rev. *Plant Pathol.* **55**, 3271.

Teakle, D. S. (1969). Fungi as vectors and hosts of viruses. *In* "Viruses, Vectors, and Vegetation" (K. Maramorosch, ed.), pp. 23–54. Wiley (Interscience), New York.

Teri, J. M., Thruston, H. D., and Plaisted, R. L. (1977). The effect of potato virus X on the yield of the potato variety Hudson. *Am. Potato J.* **54**, 271 (Abstr.).

Thanassoulopoulos, C. C., and Hooker, W. J. (1968). Factors influencing infection of field grown potato by *Verticillium albo-atrum*. *Am. Potato J.* **45**, 203–216.

Thirumalachar, M. J., and O'Brien, M. J. (1977). Suppression of charcoal rot in potato with a bacterial antagonist. *Plant Dis. Rep.* **61**, 543–546.

Thomas, D. G. (1946). Powdery mildew of potato. *Nature (London)* **158**, 417–418.

Thompson, A. D. (1959). Potato viruses A and S in New Zealand. *N.Z. J. Agric. Res.* **2**, 702–706.

Thornberry, H. H. (1966). Index of plant virus diseases. *U.S., Dep. Agric., Handb.* **307**.

Thorne, G. (1945). *Ditylenchus destructor* n. sp., the potato rot nematode, and *Ditylenchus dipsaci* (Kühn, 1857) Filipjev 1936, the teasel nematode (Nematoda: Tylenchidae). *Proc. Helminthol. Soc. Wash.* **12**, 27–34.

Thurston, H. W., Leach, J. G., and Wilson, J. D. (1948). Chromates as potato fungicides. *Am. Potato J.* **25**, 406–410.

Todd, J. M. (1954). Potato wildings and witches' broom in Scotland. *Plant Pathol.* **3**, 17–20.

Turnquist, O. C. (1976). Production of certified seed potatoes by varieties, 1974–75. *Am. Potato J.* **53**, 293–304.

Twomey, J. A., Akeley, R. V., Knutson, K. W., and Workman, M. (1968). Oromonte: A chipping potato for Colorado. *Am. Potato J.* **45**, 297–299.

Van Denburgh, R. W., Hiller, L. K., and Koller, D. C. (1979). Cool temperature induction of brown center in 'Russet Burbank' potatoes. *HortScience* **14**, 259–260.

Van Denburgh, R. W., Hiller, L. K., and Koller, D. C. (1980). The effect of soil temperatures on brown center development in potatoes. *Am. Potato J.* **57**, 371–375.

Van der Plank, J. E. (1936). Internal brown fleck; a phosphorous-deficiency disease of potatoes grown on acid soils (Union South Africa). *Dep. Agric. and For. Sci. Bull.* **156**, 1–22.

Vargos, L. S., and Nielsen, L. W. (1972). *Phytophthora erythroseptica* in Peru: Its identification and pathogenesis. *Am. Potato J.* **49**, 309–320.

Vruggink, H., and Maas-Geesteranus, H. P. (1975). Serological recognition of *Erwinia carotovora* var. *atroseptica*, the causal organism of potato blackleg. *Potato Res.* **18**, 546–555.

Wade, G. C. (1949). An unusual potato rot. *J. Aust. Inst. Agric. Sci.* **15**, 42–43.

Waggoner, P. E. (1956). Variation in *Verticillium albo-atrum* from potato. *Plant Dis. Rep.* **40**, 429–431.

Walker, J. C. (1941). Disease resistance in the vegetable crops. *Bot. Rev.* **7**, 458–506.

Walker, J. C. (1952). "Diseases of Vegetable Crops." McGraw-Hill, New York.

Walker, J. C. (1953). Disease resistance in vegetable crops. II. *Bot. Rev.* **19**, 606–643.

Walker, J. C. (1957). "Plant Pathology," 2nd ed. McGraw-Hill, New York.

Walker, J. C. (1965). Disease resistance in the vegetable crops. III. *Bot. Rev.* **31**, 331–380.

Walker, J. C. (1969). "Plant Pathology," 3rd ed. McGraw-Hill, New York.

Walker, J. C., and Larson, R. H. (1939). Yellow dwarf of potato in Wisconsin. *J. Agric. Res.* **59**, 259–280.

Walkinshaw, C. H., and Larson, R. H. (1959). Corky ring spot of potato, a soil-borne virus disease. *Res. Bull.—Wis., Agric. Exp. Stn.* **217**.

Wallin, J. R. (1962). Summary of recent progress in predicting late blight epidemics in the United States and Canada. *Am. Potato J.* **39**, 306–312.

Webb, R. E., and Buck, R. W. (1955). A diagnostic host for potato virus A. *Am. Potato J.* **32**, 248–253.

Webb, R. E., and Schultz, E. S. (1958a). Transmission and physical properties of a virus isolated from plants grown from corky ringspot-affected tubers. *Am. Potato J.* **35**, 448 (Abstr.).

Webb, R. E., and Schultz, E. S. (1958b). Tuber net necrosis in relation to the time of infection with potato leafroll virus. *Am. Potato J.* **35**, 728–729.

Webb, R. E., Larson, R. H., and Walker, J. C. (1952). Relationship of potato leaf roll virus strains. *Res. Bull.—Wis., Agric. Exp. Stn.* **178**.

Weber, G. F. (1973). "Bacterial and Fungal Diseases of Plants in the Tropics." Univ. of Florida Press, Gainesville.

Weber, W. W. (1947). Russet Sebago potato developed. *Wis., Agric. Exp. Stn., Bull.* **472**.

Weigle, J. L., Kehr, A. E., Akeley, R. V., and Horton, J. C. (1968). Chieftain: a red-skinned potato with attractive appearance and broad adaptability. *Am. Potato J.* **45**, 293–296.

Weingartner, D. P. (1977). Development of late blight forecasting and spray advisories for potatoes in northeast Florida *Am. Potato J.* **54**, 507–508 (Abstr.).

Weingartner, D. P., Shumaker, J. R., and Smart, G. C. Jr. (1977). Corky ringspot control. Hastings, Fla. Agr. Res. Rep. PR 77-4.

Weinhold, A. R., and Bowman, T. (1977). Relationship between Rhizoctonia disease of potato and tuber yield. *Abst., 69th Annu. Meet. Am. Phytopath. Soc.* p. 119.

Weinhold, A. R., Oswald, J. W., Bowman, T., Bishop, J., and Wright, D. (1964a). Influence of green manures and crop rotation on common scab of potato. *Am. Potato J.* **41**, 265–273.

Weinhold, A. R., Bowman, T., and Bishop, J. (1964b). Urea-formaldehyde for the control of common scab of potato. *Am. Potato J.* **41**, 311–321.

Weinhold, A. R., Bowman, T., and Hall, D. H. (1978). Rhizoctonia disease of potato in California. *Am. Potato J.* **55**, 56–57.

Wellman, F. L. (1972). "Tropical American Plant Disease." Scarecrow Press, Metuchen, New Jersey.

Westcott, C. (1971). "Plant Disease Handbook," 3rd ed. Van Nostrand- Reinhold, New York.

Western, J. H., ed. (1971). "Diseases of Crop Plants." Wiley, New York.

Wetter, C. (1961). Über die X-virus-Immunitat der Kartoffelsorte Saphir. [On the virus X immunity of the Saphir potato variety.] *Phytopathol. Z.* **41**, 265–270.

Wetter, C., and Brandes, J. (1956). Untersuchungen über das Kartoffel- S-Virus. *Phytopathol. Z.* **26**, 81–92.

Wetter, C., and Volk, J. (1960). Versuche zur Übertragung der Kartoffelviren M und S durch *Myzus persicae* Sulz. [Experiments on the transmission of potato viruses M and S by *M. persicae.] Eur. Potato J.* **3**, 158–163.

Wheeler, E. J., and Akeley, R. V. (1961). The Onaway potato, an early variety. *Am. Potato J.* **38**, 353–355.

Wheeler, E. J., Stevenson, F. J., and Moore, H. C. (1944). The Menominee potato: A new variety resistant to common scab and late blight. *Am. Potato J.* **21**, 305–311.

Whitehead, R., McIntosh, T. P., and Findlay, W. M. (1953). "The Potato in Health and Disease," 3rd ed. Oliver & Boyd, London.

Willis, C. B., and Larson, R. H. (1960). A new host for potato virus X in the Leguminosae. *Phytopathology* **50**, 659.

Wilson, J. D. (1968). Soil fumigation. *Am. Potato J.* **45**, 414–426.

Winstead, N. N., Wells, J. C., and Sasser, J. N. (1958). Root knot control in vegetable crops using D-D and EDB with and without vermiculite as a carrier. *Plant Dis. Rep.* **42**, 180–183.

Wolcott, A. R. (1956). Varietal response to climate and culture as related to the development of internal browning in potato tubers. *Diss. Abstr.* **16**, 844–845.

Wolcott, A. R., and Ellis, N. K. (1956). Associated forms of internal browning of potato tubers in northern Indiana. *Am. Potato J.* **32**, 343–352.

Wolcott, A. R., and Ellis, N. K. (1959). Internal browning of potato tubers: Varietal susceptibility as related to weather and cultural practices. *Am. Potato J.* **36**, 394–403.

Wollenweber, H. W., and Reinking, O. A. (1935). "Die Fusarien." P. Parey, Berlin.

Wooliams, G. E. (1966). Host range and symptomatology of *Verticillium dahliae* in economic, weed, and native plants in interior British Columbia. *Canad. J. Pl. Sci.* **46**, 661–669.

Wortley, E. J. (1915). The transmission of potato mosaic through the tuber. *Science* **42**, 460–461.

Wright, N. S. (1947). A Stemphylium leaf spot on potatoes in British Columbia. *Sci. Agric.* **27**, 130–135.

Wright, N. S. (1950). Witches' broom of potato in British Columbia. *Proc. Can. Phytopathol. Soc.* **17**, 11 (Abstr.).

Wright, N. S. (1952). Studies on the witches' broom virus disease of potatoes in British Columbia. *Can. J. Bot.* **30**, 735–742.

Wright, N. S. (1954). The witches' broom virus disease of potatoes. *Am. Potato J.* **31**, 159–164.

Wright, N. S. (1968). Evaluation of Terraclor and Terraclor Super-X for the control of Rhizoctonia on potato in British Columbia. *Can. Plant Dis. Surv.* **48**, 77–81.

Wright, N. S. (1970). Combined effects of potato viruses X and S on yield of Netted Gem and White Rose potatoes. *Am. Potato J.* **47**, 475–478.

Wright, N. S. (1974). Retention of infectious potato virus X on common surfaces. *Am. Potato J.* **51**, 251–253.

Wright, N. S. (1977). The effect of separate infections by potato viruses X and S on Netted Gem potato. *Am. Potato J.* **54**, 147.

Wright, N. S., and Cole, E. F. (1976). Control of potato virus X and potato virus S in two seed growing areas of British Columbia. *Am. Potato J.* **53**, 365.

Wright, N. S., and Hughes, E. C. (1964). Effect of defoliation date on yield and leafroll incidence of potato. *Am. Potato J.* **41**, 83–91.

Wright, N. S., Mellor, F. C., Cole, E. F., and Hyams, C. M. (1977). Control of PVX and PVS in seed potatoes. *Can. Agric.* **22**, 14–16.

Yang, T. C., and Hooker, W. J. (1977). Albinism of potato spindle tuber viroid-infected Rutgers tomato in continuous light. *Am. Potato J.* **54**, 519–530.

Young, D. A., and Davies, H. T. (1975). Belleisle: A new maincrop variety with excellent table quality and bruise resistance. *Am. Potato J.* **52**, 51–55.

Young, L. C., and Young, D. A. (1958). A study of the multi-genic type of resistance to *Phytophthora infestans*. *Am. Potato J.* **35**, 449 (Abstr.).

Young, L. C., Davies, H. T., Young, D. A., and Munro, J. (1960). Fundy: A new smooth, early maturing variety of potato. *Am. Potato J.* **37**, 274–277.

Young, L. C., Davies, H. T., Lawrence, C. H., and Hodgson, W. A. (1962). Avon: A new potato variety with moderate resistance to common scab, excellent cooking quality, and ability to size tubers early. *Am. Potato J.* **39**, 363–367.

Young, R. A. (1956). Control of the early maturity disease of potatoes by soil treatment with Vapam. *Plant Dis. Rep.* **40**, 781–782.

Young, R. A., and Tolmsoff, W. T. (1957). Cultural and chemical treatments for control of the early maturity disease of potatoes. *Phytopathology* **47**, 38 (Abstr.).

Zachmann, R., and Baumann, D. (1975). *Thecaphora solani* on potatoes in Peru: Present distribution and varietal resistance. *Plant Dis. Rep.* **59**, 928–931.

Zimmermann-Griess, S. (1947). Internal rust spot of potatoes. *Palest. J. Bot., Jerus. Ser.* **6**, 174–180.

Index

227